Dear Phillip,
Happy Birthday, you old fart!
Hope this is as interesting
as it looks, and of course,
you have to lend it to me
when you're finished!

~ Love Kelly

Wily Violets
&
Underground
Orchids

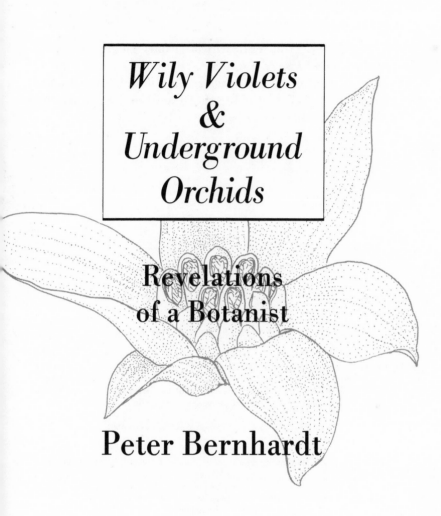

Wily Violets & Underground Orchids

Revelations of a Botanist

Peter Bernhardt

William Morrow and Company, Inc.
New York

AN EARLIER VERSION of chapter 14 appeared previously in *Natural History*. Previous versions of chapters 11 and 12, were published in *Good Gardening: Australia and New Zealand*. Chapters 1–3, 5–7, 9, 10, 13, 15–18 are based on articles that appeared originally in *Garden*.

Library of Congress Cataloging-in-Publication Data

Bernhardt, Peter, 1952—
Wily violets & underground orchids : revelations of a botanist / Peter Bernhardt.
p.   cm.
Bibliography: p.
Includes index.
ISBN 0-688-08350-1
1. Plants.   2. Botany.   I. Title.   II. Title. Wily violets and underground orchids.
QK81.B45   1989
581—dc19   89-30538
                                                                CIP

Printed in the United States of America

First Edition

1   2   3   4   5   6   7   8   9   10

BOOK DESIGN BY WILLIAM McCARTHY

To the memory of Sarah Samuels—
who knew that her grandson would make his way

Foreword

I FIRST MET Peter Bernhardt fifteen years ago when he was an inquisitive, literate Peace Corps volunteer in El Salvador trying to make sense out of the bewildering diversity of tropical plants that he was encountering for the first time, and coordinating the activities associated with the Universidad de El Salvador's herbarium and collection of living orchids. His early letters to me mark him as having exceptional insight into the biology of plants and the often curious adaptations that suit them for life in a tropical environment.

Through correspondence and occasional visits with him, I followed Peter as he moved to Australia, where he enrolled as a graduate student and research fellow under the direction of Professor Bruce Knox at the University of Melbourne. Bruce Knox is one of the most gifted floral biologists in the world; under his direction, Bernhardt's skill in interpreting the ways in which flowers function improved, while he maintained both a great enthusiasm for plants and an ability to communicate that enthusiasm to other people. Since the research group in plant cell biology at Melbourne is operating at the very cutting edge of floral biology, Bernhardt mastered the most advanced techniques and communicated on a daily basis with scientists who were making discoveries that laid the foundation for a new understanding of pollen and the way in which it functions.

I was extremely pleased when Peter Bernhardt came to St. Louis University several years ago, and hastened to offer him a research assistantship at the Missouri Botanical Garden. Here I often find him outside my office, dissecting orchid flowers or squashing iris pistils as he pursues the investigations that form the basis of this book and his scholarly work in general. Always a stimulating discussant on floral evolution and ecology, Bernhardt has enhanced the abilities of students and professionals alike to interpret the functioning of flowers.

Why does any of this matter? Our relationship with the plant world is based on reciprocity. Plants are all around us, and through their ability to capture some of the sun's energy and make it available in chemical form for biological processes, they make all life possible. But this is only one side of the equation. Plants are immobile, incapable of moving from place to place in search of food or mates. The food they produce themselves. The mates, on the other hand, are found by their control of the activities of insects and other animals. Plants have developed the ability to direct animal activities through such adaptations like flowers and edible or adherent fruits. In this sense, flowers are much more than just colorful and often delightfully scented structures; they are organs designed to manipulate us animals into helping plants propagate. Our interactions with plants are remarkably two-sided—without them, life as we know it would not exist; and without living, mobile animals, many plants would find it very difficult to outbreed successfully, or to disperse their seeds.

By understanding more about the biology of flowers, therefore, we can appreciate plants better and enrich our interactions with them, whether as gardeners, agriculturists, or simply as lovers of beautiful objects. The everyday flowers around us—oaks, chrysanthemums, irises, and azaleas that fill our gardens, woods, and prairies—and also the exotic and often bizarre plants that we encounter in greenhouses or on occasional trips to warmer parts of the globe—become more interesting and vital the better we know them.

Drawing on his own rich experience and the observations of hundreds of other scientists, Peter Bernhardt presents us with a rich banquet of floral adaptations. Peter sets the stage by contrasting the very different, yet still fundamentally similar, lives of plants in the tropics, the seasonally dry Australian forests, and the prairies of North America; then he moves on to outline the ways in which plants and their pollinators relate to one another. We meet Australian possums, Mediterranean scarab beetles, and primitive moths from the South Pacific, as well as the more familiar bees, flies, and butterflies. Peter then uses the ways in which flowers function today to illuminate theories about their evolution. In subsequent sections of the book, the lives of a number of different kinds of plants are treated in delightful and satisfying detail, from violets to giant Amazonian water lilies to tropical air plants. Finally, Peter deals with that most bizarre of all plant families, the orchids. Here the reader is led to understand what really sets these plants apart from others, and to speculate as to why there are so many different kinds of orchids in the world—as many as one in ten of all kinds of plants is an orchid. Because they are numerous, because most are tropical, and because most have unique patterns of floral adaptation and often incredible beauty, orchids are of special interest to us.

I believe that any reader will be stimulated and intellectually enriched by this book, written as it is in a lively and engaging style that clearly and precisely documents the adaptations that have, over the course of time, made plants what they are. No one can come away from these essays without a clear sense of the vitality of botany and the special interest that plants hold for us all. Their flowers have coevolved with animals, including ourselves, and they form a living context for human life. When our earliest ancestors appeared on the savannas of Africa, they survived because of their ability to use plants—their fruits, leaves, wood, scents—for their own purposes. Our brains evolved largely in relation to our perceptions of the natural world, and even today, when so many of us live in cities and are so widely separated from nature, we fill our gardens and our homes with plants because we delight in them. By understanding them better, we can appreciate them even more, and also gain a sense of the ways in which scientists of many countries are contributing to the development of knowledge about them. This knowledge is constantly being applied to the improvement of cultivated plants and to the better utilization of natural resources, and will be even more important in the world of the future than it is today.

Accentuating the urgency of our considerations is the fact that the growth of our own population is threatening the continued existence of plants throughout the world. For example, a survey recently conducted by the Center for Plant Conservation revealed that nearly 300 species of plants in the continental United States are in danger of extinction by the mid-1990s unless specific steps are taken to preserve them. Throughout the world, the activities of a human population that has doubled since the 1950s, coupled with widespread poverty and malnutrition, are threatening to eliminate perhaps a quarter of all species of plants on Earth. As many as 60,000 species, therefore, may vanish by early in the next century unless we take steps to locate and preserve them. Viewed against this background, the kinds of studies that are discussed so eloquently by Peter Bernhardt in this book are essential both to enhance our appreciation of plants and therefore our desire to preserve them, and also to lay the foundation for scientific advances in that understanding. In a very real sense, plants are the only sustainable resource that we have: cherishing it, preserving it, and learning to use this resource more efficiently are necessary ingredients for human prosperity in the future.

Peter H. Raven
Director, Missouri Botanical Garden
St. Louis, Missouri

Contents

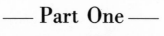

— Part One —

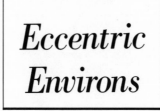

Eccentric Environs

Chapter 1

Trees of Two Seasons

Nessus had not yet reached the other side
   when we were on our way into a forest
   that was not marked by any path at all.

No green leaves, but rather black in color,
   no smooth branches, but twisted and
   entangled
   no fruit, but thorns of poison bloomed
instead.

—*Dante Alighieri*
(translation by Mark Musa)

ONE SEPTEMBER FIFTEEN years ago, when I first assumed my Peace Corps appointment in El Salvador, I accepted an invitation to spend a weekend on the country's Pacific coast. We were driven down to the beach colony near the town of La Libertad. There I found the coastal forest in all its botany-textbook splendor. There were huge, buttressed native figs, stinging nettles that had taken the form of wiry saplings, and stands of mango, coconut, guava, limes, and cashew that had escaped from cultivation but were offering their tasty, feral fruits to poor campesinos and wealthy holiday makers. Excluding the brilliant tints on the wings of the large butterflies, the dominant natural color seemed to be shades of glossy green.

Four months later I traveled to the coast for another weekend by the sea. As the bus drew closer to the forests of the Deininger Park Preserve, I noticed that the landscape had undergone a spectacular metamorphosis. The vegetation now resembled a hardwood forest in the American Northeast. Many trees were completely bare of leaves, while others wore yellow, purple, and red robes, reminiscent of groves of autumnal maples, beeches, and oaks.

Seasonality is definitely part of the tropics, and nowhere so apparent as along the Central American coasts, lowland plains, and inland valleys from sea level to about eighteen hundred feet up the volcanic slopes. These cyclically dry forests are adapted to a climate that provides rather monotonous daily temperatures but alternates a long wet season with a long dry season. At one time, these forests occurred extensively over the Central American lowlands, but they have been reduced severely, and surviving habitats have been disrupted through a combination of burgeoning populations, the proliferation of resort areas, and the spread of cash crops, such as sugarcane and cotton. Remnants survive from Mexico south to Panama, and they can be very impressive. Over a hundred different species of woody plants may be crowded into just a few square miles of lowland forest. Tree species that extend their branches to form the overlapping, shady canopy may stand more than 95 feet high. The comparatively smaller species that make up the second living layer, or understory, tend to have more slender and crooked trunks and only reach a pitiful 65 feet, at best.

The many species that comprise the lowland forest are mastered by water. If a tree is to live long enough to reproduce, it must live through alternating periods when water is either too plentiful or absent entirely. The two seasons tend to vary in length according to the position of countries in relation to the equator. Generally speaking, the closer to the equator, the longer the rainy season. Rain may fall every day from late May through early November in El Salvador. Conversely, there is an annual drought from mid-November until mid-May. (See color plates 1 and 2.)

During the rainy season, deluges occur almost daily from late in the afternoon until early in the evening. From June through July, El Salvador experiences storms known as temporales that last several days to a week. Lowland forests in El Salvador may receive 60 to 80 inches of rain over five months. That does not mean that the forest floor becomes waterlogged or sodden. Tropical topsoils remain rather thin, since the heavy rains wash the absorbent humus layer away almost as soon as it forms. Runoff tends to be rapid in these forests, so proportionately little water stays in the hard, rocky substrates that remain—reddish laterites or black montmorillonite clays. I have walked along forest paths less than an hour after a downpour without need of boots. Despite the calendar, Central Americans refer to the rainy season as "winter." It's an unexpected choice of words. The rain tends to be lukewarm and can make a pleasant and soothing shower. This is also the season in which forest trees leaf out in full.

The dry season is the Latin American "summer." Leaves change from a necessity to an expensive luxury during this period. Not only do leaves require a lot of water to stay alive, they also increase a tree's loss of stored water as vapor escapes through the tiny porelike openings (stomates) that dot the underside of most green foliage. Unless trees grow near a wide river or swamp, they have no long-term reserves.

During my hikes through the Deininger Preserve and the lowland forests of Tikal, in Guatemala, I couldn't help but notice how noisy the forest seemed in March or April. The branches were so dry and there was so much crunchy leaf litter on the forest floor that it seemed anything larger than a tarantula was making an ungodly racket. Dawn was especially loud as troupes of animals made their daily rounds. A family of howler monkeys would crash through the branches only to be followed by a small but equally indelicate flock of toucans searching for insects disturbed by the movement of the primates. On the ground you could see bands of racoonlike coatimundis scraping the leaf litter for fallen fruit and more bugs. So much for silent, cautious wildlife.

Natural selection favors those plant species that find a way to beat the fatal thirst of the dry season, thus encouraging the cyclical "striptease." Some tropical ecologists prefer to call these habitats xeric caducifolious forests, which literally means "leaf-dropper during dryness." Leaf loss occurs at different rates for different trees and may depend on how well individual species can store water in their foliage or whether local populations are growing near standing bodies of fresh water. Some low-altitude trees kept moist artificially will drop their leaves at the appointed time but then will replace the lost crop of foliage with fresh new leaves less than a month later. Other species will keep their leaves year-round if they are watered regularly.

Where cyclical leaf drop is common, the foliage turns yellow and brittle as chlorophyll breaks down and excess water is transported back into the limbs. Then the cells at the base of the leafstalk (petiole) form an abscission zone, and the leaf drops. An impermeable, rubbery layer composed of a substance called suberin forms over the leaf scar. By the time all the leaves have fallen, the limbs are properly sealed from the sun.

When the dry season comes to a close, trees begin to produce new leaves at least a month before the first continuous rains, which suggests that some species cycle to light rather than to soil moisture. Trees need to produce leaves as early as possible if they are to compete successfully for available sunlight with the trees around them. A species that waited until water was in constant supply would find itself outshaded by its peers and thus unable to produce enough reserves in time to withstand the next dry season. So, photoperiod is a more dependable cue than the first drizzling rains, in most cases.

A few species do stay green year-round, though, despite the severity of the dry season. Their limbs and/or foliage are modified to an extreme. Large, shrubby prickly pears *(Opuntia)* grow throughout some of these lowland regions and store water in their succulent yet photosynthetic stem pads, like most cacti. However, their true leaves are reduced to ephemeral stubs with pitted stomates that do not release or absorb gases. A few trees, such as the devil's apple *(Clusia),* actually retain a dense foliage all year long, but their leaves are considerably shorter than those of the neighboring species that produce a lush, new canopy each rainy season. Each leaf of a devil's apple is protected by a dense, waxy coat and has only a few functional stomates on its underside to help slow the inevitable loss of vapor.

Leaf drop offers a tropical tree a secondary benefit. Because winter isn't freezing cold in the tropics, the life span of insects such as large beetles and butterflies may be extended several months. They continue to reproduce in surroundings that are warm year-round; and that can provide new generations of munching larvae with a steady source of food. When trees lose their leaves, however, it becomes insect winter. Populations crash, and the bugs survive as dormant eggs or pupae. The few tree species that retain their leaves all year remain remarkably free of insect damage. Most, like the devil's apple, have evolved mechanisms—such as extremely acidic leaves—that discourage all but the most specialized insects.

Of course, even during the rainy season, the lush canopy is definitely not a natural salad bar for hordes of caterpillars. The vast majority of trees in these lowland tropics arm their leaves, branches, and seeds with a complex array of chemical irritants and toxins. No species of larva, no matter how hungry it is, can possibly evolve digestive defenses against all of these poisons. So, each insect species tends to confine its eating to just a few tree species.

Often they attack only certain plant organs or simply feed on plant tissues that are still young and have not yet developed their full toxic strength. The caterpillars of the tropical silk moths that feed on the leaves of a hog plum *(Spondias)* have about as much of an appetite for the foliage of a coral tree *(Erythrina)* as Macbeth had for the contents of the witches' cauldron.

The biochemical arsenal can work in a variety of ways. Some tree organs are "spiked" with pockets of tissues called lactifers and resin ducts. When the mouthparts of an insect puncture or chew into these tissues, they release gluey fluids that harden upon contact with the air. Latex and resins can gum up insect mandibles, but they also contain small quantities of volatile substances, so that the overall effect must be like swallowing rubber cement flavored with chili oil.

Fatal poisons work at a much more subtle level. The nitrogen-rich class of plant toxins known as alkaloids bind to animal enzymes and cause liver deterioration. Phenols and bitter tannins, stored in the water bags (vacuoles) of plant cells, can slow or prevent the process of digestion. Ten percent of the dry weight of the seeds of some of the *Dioclea* and *Canavalia* trees is an unusual amino acid called canavanine. Insect bodies mistake canavanine for the essential amino acid arginine, with fatal consequences. Daniel Janzen, the great tropical ecologist and evolutionary biologist, summed up this forest of Borgias when he wrote, "The world is not coloured green to the herbivore's eyes, but rather is painted morphine, L-DOPA, calcium oxalate, cannabinol, caffeine, mustard oil, strychnine, rotenone, etc." It is no coincidence that both the drugs for humans and many organic insecticides and rodent poisons have been derived from plants of the lowland forests of the tropical Americas.

During the dry season, the absence of green leaves does not mean that a lowland tree has entered a state of dormancy. Leafless, the gumbo limbo or jiote *(Bursera simaruba)* continues photosynthesizing at a limited level, since functional chloroplasts live just under the bark. However, the dry season is the time of year in which lowland trees are most likely to express their sexuality.

Flowering is possible because flowers are not the heavy users of water that leaves are. Furthermore, the blossoms lack functional stomates, and the life span of a flower is usually measured in days, not months.

The manufacture of flowers, fruit, and seed may not be expensive compared with the water used by leaves, but it is very costly in terms of energy and stored nutrients. Trees that have not had a successful rainy season or have been weakened by predators or fire will forgo flowering for one dry season or more. Even in some of the most productive forests, individual trees of certain species fail to flower annually. The almendro macho *(Andira inermis)* is a woody member of the bean family that usually flowers once every two years. A common kapok tree *(Ceiba pentandra)* must often wait five to ten years between flowering periods. Field studies completed in the lowlands of Panama note that there are a few tree species that grow for a couple of decades, flower once, then die. *Tachigalia versicolor,* a woody relative of our sennas and partridge peas, is one such species that turns reproduction into a Pyrrhic victory.

All the tree species in a lowland forest do not burst into bloom at the same moment, although the broadest peak occurs from January to March, and then some more species "pop open" by mid or late April, usually just before the rains come. Salvadorans know the white-and-gold-spotted flowers of the frangipani *(Plumeria rubra)* as the flor de Mayo because the first buds open before the first tentative storms of May.

The dominant trees of our northern American forests are likely to produce reduced flowering branches known as catkins (see chapter 5) with inconspicuous flowers that shed their pollen on the breeze. Since the lowland tropical forests are such a crazy quilt of plant life, members of the same species do not often have the luxury of forming densely populated groves within a single area. The odds are against air currents removing a pollen grain from one flower and depositing it on the appropriate pistil of a second tree of the same species because two individuals of the same species may be growing many yards apart. This is why most trees of the lowland tropics are pollinated by animals, and that includes species with rather tiny flowers, even figs and palms. Animals seek out edible rewards, such as nectar or pollen, and are entrusted with the tree sperm locked inside pollen grains. In the evolution of lowland tree flowers, a wide selection of potential pollinators have been utilized, including bees, butterflies and moths, beetles, flies, birds, and bats. Fifty-three species of hummingbirds are found throughout Mexico and northern Central America alone.

Lowland trees manipulate the foraging behavior of their pollinators by the way in which they present their flowers. The flowering behavior of the tree species may be divided into two rather broad categories. There are species in which a tree bears thousands of flower buds that burst into bloom simultaneously and remain in flower for a short time. This flowering pattern is charmingly named the "Big Bang" strategy, and it is particularly common in members of the tropical trumpet-vine trees (family Bignoniaceae). Bare branches of the cortes *(Tabebuia ochracea)* turn into golden torches for just a few weeks of the dry season, but they attract a wide variety of opportunistic pollinators searching for a quick energy fix or a temporary, but generous, source of pollen to feed to their young, in the case of the larger tropical bees. The densely packed clusters of flowers are crowded by insect visitors and nectar-foraging birds.

In contrast, there are species on which only a few large flowers are open at any one time. The same tree may stay in flower for months as it unobtrusively replaces its spent flowers with freshly opened buds every couple of days. Coral trees *(Erythrina)* and the gourd tree *(Crescentia alata)* tend to exhibit this "Steady State" strategy, which caters to more specialized nectar feeders willing to "trapline" for floral rewards. That means that these animals remember small, but concentrated and self-replenishing sources of food and return to them day after day, even though they may have to visit a few trees dispersed over a wide area. Trees exploiting the Steady State mode of presentation are pollinated by comparatively few species of animals that tend to be strong fliers and have long tongues, such as bats, nocturnal sphinx moths, hummingbirds, tropical bees such as *Melipona* (the wild honey source of the Mayans) or the euglossine bees.

Steady supplies of nectar and/or pollen can make the difference between survival and loss of a generation during the dry season. Larger pollinators may protect their food sources instead of proscribed territories. Tropical entomologists have commented on how some worker bees attempt to chase other bees away from mass flowering caraos *(Cassia),* major producers of pollen for bee larvae. Native coral trees are often grown as windbreaks around lowland plantations and form artificially rich sources of nectar. As trains roll past these plantations, you can see a hummingbird every few feet, perched on the barbed wire that strings the line of trees together. Each bird is prepared to defend the nearest branches of scarlet, tubular flowers.

Once the pistils are fertilized, seed and fruit development begins. Some tree species have ripe fruits long before the end of the dry season begins, whereas others delay the complete maturation of the fruit wall until many months into the dry season. Delayed ripening has some advantages. Once the seeds reach maturity and drop from the parent tree, they will fall onto ground that is both wet and soft so that they can sprout immediately and anchor themselves in the ground.

There is a price to be paid for delayed ripening, however. The longer it takes for seeds to develop and be released by their parent tree, the greater the chance of predation. Although the seeds of many lowland trees are quite toxic, there are always a number of specialized insects that remain intrepid predators. They may be bruchid beetles, weevils, coreid bugs, or pyralid moths but their mode of attack is pretty much the same. The pregnant adult lands on the developing fruit of her choice. She lays one or more eggs and then flies off. The hatching larvae will burrow through the tissues of the fruit wall, and each grub will eat one or more seeds until it is large enough to pupate.

In some ways the seeds of lowland trees represent the weakest link in the life cycle of their species; their mortality cannot be underestimated. The pods of the raintree *(Pithecollobium saman)* are very susceptible to the grubs of bruchid beetles; 50 to 70 percent of the seeds produced each year fail to survive. Some trees suffer attacks from several different kinds of seed predators. When the seeds of achiote *(Bixa orellana)* are young and soft, many are devoured by the nymphs of coreid bugs. Those seeds that survive to reach their full size and hardness may still succumb to a second hungry wave of bruchid grubs.

Therefore, once a seed reaches maturity, it cannot remain long on the parent tree, and it would still be an easy target if it just fell onto a pile of seeds on the ground, so the lowland species have devised dazzling ways to disperse their seeds. Some species shed their seeds on the wind. When the large pods or capsules are ripe, they split open to release seeds wearing little glider wings or propellers. Others emerge from the interior of the fruit wrapped in woolly threads known as kapok. Kapok acts as a parachute for the hard, wingless seeds of the balsa *(Ochroma lagopus)* and the silk tree *(Cochlospermum vitifolium)*. These fibers have other uses as well. You can often see Central American women harvesting the open pods of silk trees from May through July, gathering the soft kapok with which to stuff pillows.

Other species depend on animals to transport their seeds. Hard, indigestible seeds are attached to colorful and tasty bits of food. Fruit tissues vary in nutritional quality. Some consist primarily of simple sugars, such as the lowland figs, while others may be rich in oils and some amino acids, such as some of the many wild relatives in the family of the cultivated avocado (Lauraceae). Fruits of the hog plum *(Spondias mombin)*, agujon *(Brumelia persimilis)*, and nance *(Byrsonima crassifolia)* offer pungent-smelling, sugary fruits, which are devoured by bats, coatimundis, and monkeys that pluck the rewards right off the twigs. The acajou *(Anacardium excelsum)* is a tall tropical tree related to our own sumacs *(Rhus)*. Its fruit is a hard-shelled nut, but it still produces a juicy reward for bats. When the nut is ready for dispersal, the little stalk, or receptacle, that connects it to its twig turns soft and sweet. Bats carry off the stalks at night and eat them in flight. The hard nuts plummet to the ground.

Fruits that just drop from their branches onto the ground also stand a good chance of further dispersal if they attract peccaries, deer, coyotes, tapirs, and huge rodents like the pacas and agoutis. The agouti may be especially useful in the establishment of new forests, since it buries larger seeds as a squirrel would.

During the dry season, fruit may become the most important component in the diet of birds such as toucans, some parrots, orioles, and cotingas. A number of trees' seeds are dispersed almost exclusively by birds. When the fruits are ripe, they split open to reveal a Whitman's Sampler of juicy or oily seeds. Each seed wears a fleshy coat or cap (aril) that is bright red or orange to attract birds. Birds eat the arils, then drop or defecate the hard seeds intact. *Stemmadenia* is a genus of tree that belongs to the oleander family (Apocynaceae). The fruits tend to be large, oblong, and paired, so the trees receive rather rude names in Spanish, such as huevos de caballo (horse's eggs), or cojon de puerco (pig's testicle). The large, fleshy seeds of *Stemmadenia donnell-smithii* are carried off by at least twenty-two species of birds in Costa Rica.

The tropical lowland forest has even odder trees, which apparently once offered fruit to large animals that have been extinct for more than ten thousand years. Species like the gourd tree *(Crescentia alata)*, the stinking toe or guapinol *(Hymenaea courbaril)*, the guanacaste or earfruit *(Enterolobium cyclocarpum)*, and several members of the custard apple family *(Annona* and *Sapranthus)* produce extremely large fruits with thick, hard or leathery rinds and a sweet, mushy pulp that may not ripen until the heavy fruit falls off the tree. These fruits never split open (dehisce) along natural sutures. They must be broken apart by hard hooves or large teeth, otherwise the seeds will not be freed to sprout. Peccaries, tapirs, and agoutis can gnaw their way into a few fruits and may disseminate a fraction of the total seed crop. Ironically, introduced horses and cows have much more success in smashing rinds and eating the pulp, and their dung gives rise to slender new saplings in old pastures.

Why should animals introduced by Spanish colonists spread certain seeds more effectively than native mammals? Daniel Janzen and some of his co-workers contend that tree species bearing large, pulpy fruits with tough rinds may have ancestors that relied on large creatures themselves for seed dispersal. The fossil evidence indicates that the New World tropics once supported quite a different cast of land mammals. This included horselike species, which became extinct long before the arrival of the conquistadors, and some huge vegetarians, such as land sloths and the elephantine gomphotheres (related to the mastodons). As these mammals vanished during the Pleistocene, some perhaps as recently as only ten thousand years ago, tree species that did not become extinct survived because smaller mammals such as agoutis and tapirs carried on the seeding process, although at a much more limited level. For the past five centuries, introduced cows and European horses have, in part, filled a niche vacant for millennia. It may be impossible ever to prove or reject this "gomphothere fruit" theory, but it certainly cannot be dismissed. Janzen and his students are now helping to restore disrupted and overgrazed sections of lowland forest in Costa Rica by feeding Guapinol fruit to horses and letting them wander over the countryside.

In his unsettling novel *The Plumed Serpent,* D. H. Lawrence treats the dry but blooming countryside of Mexico as an exotic vista that seems vaguely sinister to expatriate Europeans. The ancient cycle of dense leaves yielding to bright flowers and strange fruits is not usually foremost in our minds when we take a holiday in Mesoamerica or the Caribbean. Nevertheless, the splendor of trees and shrubs in full bloom does coincide with Yuletide getaways, and it's one of the few tourist pleasures free from additional cost.

Chapter 2

A Child's Garden of Gumnuts

"There's one other flower in the garden that can move about like you," said the Rose. "I wonder how you do it."

—*Lewis Carroll*

AUSTRALIANS SEEM TO be on more familiar terms with their own flora and fauna than Americans. Citizens of Australia know that they share their continent with unique plants and animals, and they have found ways of reminding themselves of this uniqueness every day of their lives. A platypus swims on one side of the twenty-cent coin. The emu and the kangaroo flank the national crest. Country towns and suburbs such as Boronia, are often named after wildflowers.

This informal taxonomy has also made its way into Australia's children's literature. Books abound that delight in using eucalyptus and marsupials to create a fantasy world of marvels and quiet humor. They are all part of a trend that developed during the first two decades of this century, much of which was initiated by a single woman, May Gibbs. Her book, *The Complete Adventures of Snugglepot and Cuddlepie,* now an acknowledged children's classic, was the first to incorporate the natural history of Australian wildflowers within a fantasy framework.

British–born May Gibbs arrived in Western Australia with her parents in 1879, when she was four. Although she returned briefly to London at the turn of the century to receive formal training as an artist, her lifelong devotion was to the Australian bushland. For most of her long life she drew and wrote about a secret, enchanted community as close as the nearest forest.

Snugglepot and Cuddlepie appeared originally as three separate volumes published between 1918 and 1921. In 1946 her publishers, Angus & Robertson, combined the three into a single book, which has since gone through more than twenty hardcover printings and recently made the jump to paperback.

The first publication of *Snugglepot and Cuddlepie,* shortly after World War I, was a landmark in Australian children's fantasy. Until that time, Australian fairy stories were chained to European conventions. The old country goblins, wizards, and castles, with

Snugglepot and Cuddlepie ride away while their portraits are advertised on a treetrunk. The calligraphy parodies the designs left by tunneling larvae inside the bark of scribbly gums. Boy gumnuts wear operculum caps. Girls wear hats of eucalyptus flowers and rings of stamens for skirts.

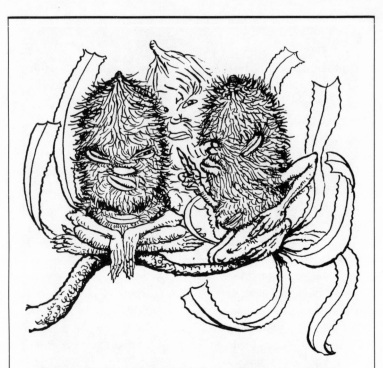

Drowsy Banksia Men spend much of their time plotting against the gumnut babies. Their meetings resemble the cob clusters on old, unburned branches of Banksia serrata. *The cobs bear several follicles, so a Banksia Man may have an extra mouth or an extremely large navel.*

plots pirated from the Brothers Grimm, were superimposed on a
sunbaked eucalyptus savanna. Historian and literary critic Hugh
Anderson wrote, "It was not until May Gibbs created her gumnut
babies and banksia men that Australian fantasy gained something
of the quality of the true fairy tale." I might add here that the
impact of *Snugglepot and Cuddlepie* in Australia parallels the
equally unprecedented success of *The Wizard of Oz* in America.

Of course, Gibbs was indebted to the classic children's authors
of the nineteenth century. Her heroine, Ragged Blossom, is probably a cousin of Hans Christian Anderson's Thumbelina. Her eucalyptus nut heroes are cast in the adventurous mold of Charles
Kingsley's waterbaby, Tom. What makes May Gibbs's work
unique, though, is that her little people symbolize the natural
history of the native flora, and they are involved in adventures that
specifically hold the interest of Australian children.

Although May Gibbs wrote and illustrated comic strips, post-
cards, and many more books of fairy tales, *Snugglepot and Cuddlepie* is far and away her best-known creation. The two sprites, or
bushbabies, as Gibbs called them, Snugglepot and his foster
brother Cuddlepie, are based on the fertilized capsules, or nuts, of
eucalyptus trees. Eucalyptus capsules do resemble old-fashioned
meat pies and lidded pots. Their best friend is an orphan girl
named Ragged Blossom, a eucalyptus flower. Since all three are
Australian fairies, small pairs of eucalyptus leaves grow on their
backs instead of wings.

Adventures begin for the young gentlemen one day when a
wise old kookaburra *(Dacelo gigas)* tells Snugglepot and Cuddlepie about Humans, who are strong as the wind, swift as the river,
cruel as the snake, and who look like pale frogs underneath their
many skins, which they can take off. Snugglepot and Cuddlepie
are so fascinated by this description that they set off in search of
Humans. This begins a series of adventures with other Australian
creatures that eventually makes the three bushbabies rich and
famous.

The plot pales before the rich and imaginative ways in which
plant anatomy has been worked into the narrative, however. Gibbs

reflects native vegetation in a funhouse mirror in a way that keeps identifiable most basic aspects of sclerophyll woodlands and shrublands, the dominant environments within the Australian landscape. *Sclerophyll* means "hard leaf," and the vast majority of Australian plants are squat, evergreen bushes or small trees that produce tough, long-lived leaves that resist burning summers and wet, cool winters.

Members of the myrtle and eucalyptus family (Myrtaceae) compose most of the canopy and a good proportion of the shrub zones. While five or six species of the genus *Eucalyptus* are naturally distributed elsewhere, over six hundred species and varieties are native to Australia. Known locally as gum trees, mallees, messmates, and bloodwoods, *Eucalyptus* species do not merely form the backbone of Australia's forests, they are the entire skeleton. Wattles *(Acacia)* and members of the macadamia nut family (chapter 6) are diverse and well distributed also.

No child reared on Snugglepot could ever fail to recognize a gum tree. Its vegetative and reproductive parts appear in the illustrations and text in ever-changing guise. The simple, curved, lance–shaped leaf is a case in point. For Gibbs it was not a mere organ of photosynthesis but, transformed into a table, ship's sail, surfboard, or artist's easel, an essential element of bushbaby civilization.

Gibbs put the fertilized eucalyptus capsule into service, as well. Gumnut laundresses wash clothes in immense potlike gumnut shells, bringing to mind Australia's old, much–hated Laundry Monday, when women boiled clothing in copper kettles. The great Australian craving for sweets, British in origin, is satisfied by the ice cream cones and cordial cups made from more conical capsules.

Gibbs's knowledge of plant anatomy was used also to satirize urban life in Australia during the early decades of this century. Children exposed to the sight gags juxtaposing botany with middle-class comforts quickly committed the identifying features of native plants to memory. Who can forget the white, pink, and scarlet eucalyptus flowers worn haughtily by Gibbs's fat koala bears? They

resemble the ludicrous hats that Sydney's matrons wore to afternoon teas or the holiday races.

Some jokes reflect real plant/insect interactions. Snugglepot
and Cuddlepie are literate if not artistic young Nuts. Throughout
the book they write and draw with a somewhat unsteady hand. This
running gag is taken directly from nature. The white trunks of trees
like *Eucalyptus sclerophylla, E. racemosa,* and *E. stigmata* are covered with wavering, squiggly black lines, giving them their common name of scribbly gums. The lines come from moth and beetle
larvae that chew permanent pathways directly under the corky
surface, leaving visible trails of darkened tissue that resemble the
unsure technique of a tot with a large crayon. Some bushbabies
keep bulldog ants *(Myrmecia)* as pets or employ them as domestics.
We now understand that many predatory ants patrol scleromorphic
foliage searching for insect prey. The vegetation benefits from the
ant/plant association, since ants kill pests or disrupt sucking and
chewing activities.

The costumes of minor characters reflect flowering and fruiting
fashions within a number of sclerophyll habitats. Eucalyptus
flowers do not have petals or sepals. The flower bud wears an
ornamented cap, known as the operculum, which pops off at maturity to expose the pistil and the brushy ring of stamens. These
opercula are favored headgear of the gumnut women. Wattle babies
are unabashedly androgynous, though, and wear unisex pajamas
composed of the massed stamens of the flowering heads of acacias.
Lilly Pilly is not only an accomplished screen actress, she is also
a parody of the lilly-pilly tree *(Acmena smithii),* another member
of the Myrtaceae. Her muff, bonnet, and skirt mimic the pale
clusters of small, sour fruits.

Every classic fantasy has its own baddies, and *Snugglepot and
Cuddlepie* is no exception. The Big Bad Banksia Men hate all Nuts,
and our heros in particular. These malicious trolls would like
nothing better than to "bunch and scrunch" Snugglepot and Cuddlepie and "root and shoot" Little Ragged Blossom. They frequently disguise themselves as ships' captains and harmless old
tramps to lure the Nuts to their doom. Undoubtedly they are May

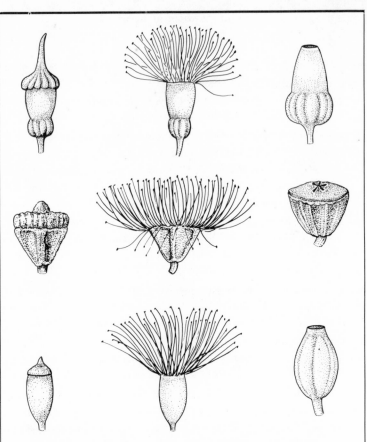

The fads and fashions of gumnut babies are based on natural variation in the flower buds (left column), flowers (middle column), and fertilized fruits (right column) of Eucalyptus *species.* Eucalyptus *flowers have no petals or sepals, so the buds wear a hard, ornamental operculum, which pops off to expose the thick brush of stamens in the open flower. From top to bottom, the species illustrated are the coral gum* (E. torquata), *the Tasmanian blue gum* (E. globulus), *and the yellow bloodwood* (E. exima). *Illustrations by J. Meyers.*

Gibbs's best floral caricatures. Every Australian with whom I've discussed the book remembers the bad banksia men first. One woman in her late twenties, now a public school teacher, admitted to having childhood nightmares in which banksia men figured prominently. When her parents drove to the beach, she would close her eyes to avoid the "evil stares" of the banksia cobs growing on trees that line many coastal roads.

Remove the clawed arms and knobby legs from a Big Bad Banksia Man and he becomes the fruiting branch, or cob, of a benign tree known as the saw banksia or red honeysuckle *(Banksia serrata)*, an important member of Australian ecosystems. The numerous whiskers covering his body are the twisted, wiry pistils of old withered flowers. The eyes, ears, nose, and "warts" are really primitive dry fruits known as follicles. In nature, every flower on the banksia's reproductive branch contains one living pistil, but only a few of these are fertilized and mature to form swollen follicles. For this reason the majority of Big Bad Banksia Men are thickly whiskered instead of thoroughly warty.

One wonders why the banksia men are so bad, considering their distinguished history. The first plants were collected by botanist Joseph Banks while he accompanied Captain James Cook on his voyage of discovery to Australia in 1770. Collecting along the coast of what would become New South Wales, Banks and his co-workers returned to the ship with so many new plants that Cook named the area Botany Bay. Years later, some of the original specimens reached Linnaeus's son in Sweden. Carl Linnaeus, Jr. named the entire genus *Banksia,* in honor of its collector. Australia continues to honor both Banks and *Banksia.* The botanist's portrait and those of the endemic plants he collected adorn one side of the Australian five-dollar bill.

While the Gibbs woodland is full of hostility between the eucalyptus children and the banksias, there is no obvious aggression in nature. I suspect that Gibbs may have invented this war after noticing that the two groups of plants do not thrive under the same conditions, although they often share overlapping distributions. This is the result of similarities between their growth rates and

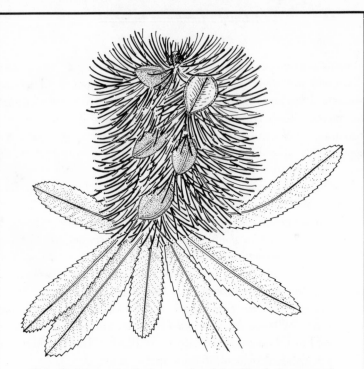

This fruiting cob of the saw banksia (Banksia serrata) *illustrates that, while a single cob may produce hundreds of individual flowers, only a few mature and become large, woody follicles. The remaining flowers wither and twist to become the sinister whiskers of the Big Bad Banksia Man. The artist has "popped" a follicle in the upper right-hand corner to show the two "lips" of each simple fruit. In nature, the two follicle valves would open only after a brushfire or a very hot summer. Illustration by J. Meyers.*

responses to cyclical bushfires. Sclerophyll leaves are rich in volatile oils, and the native vegetation experiences such hot, dry summers that bushfires are unavoidable once enough natural tinder
accumulates and lightning strikes the earth.

Severe fires kill most adult banksias outright, so they are
more likely to grow to maturity in regions that experience infrequent or weak burns. There, they form dense, tall thickets that
crowd slender eucalyptus saplings and shade out new seedlings.
Commercial eucalyptus forests in Australia are carefully burned
or cleared every few years to remove invasive banksias and other
shrubs.

In contrast, those forests that experience rapid, hot burns every
fifteen years or so tend to be dominated by a canopy of eucalyptus.
Banksias and other shrubby growths form a stunted and patchwork
layer underneath the gum trees. Many *Eucalyptus* species are
adapted to withstand summer fires. Trunks bear a thick layer of
airy, fibrous bark or hardened resin—a sort of thermal insulation.
If twig tips and leaves are killed by fire, they are replaced after the
winter rains by the buds of epicormic shoots hidden under the bark,
which grow into new limbs, regenerating lost foliage. However,
since banksias are killed by fire, they must continually regrow from
seed and therefore can't compete effectively with the overhanging,
established, self-replacing umbrella of gum leaves.

If banksias show such a low rate of survival in a country known
for its bushfires, how can any scientist suggest they are adapted to
cyclical burns? The answer lies in the one adaptive feature
banksias share with many eucalypts. Their hard, dry fruits require
the heat of bushfires to open and spread their seeds after the fire
ends and the ground has been cleared of live plants and old timber.
The two blackened valves, or "lips," on each banksia follicle gape
open slowly and disgorge one or two flakey seeds on the hot winds.
Without fires, cob numbers increase on large shrubs or trees until
they begin to resemble the ugly mobs of plotters in a Gibbs illustration.

Of course, cobs of banksias wearing weathered coats of dried
flowers are extremely flammable, and fire becomes the means

through which the Big Bad Banksia Men receive their just desserts. A lizard friend of Snugglepot and Cuddlepie becomes so thoroughly outraged by the villainy of the gumnut's enemies that he finds their hideout and throws each Banksia Man into a bonfire. This is a scenario familiar to campers in Australia's outback. To start a cookout fire to prepare the evening's "billy" (tea), nothing makes better kindling than an old banksia cob.

Snugglepot and Cuddlepie never pretends to be a field guide to the local flora, and Gibbs showed no interest in scientific prose. Frequently, though, she admonished young readers to treat bush creatures with kindness. As she once said, "I think you can influence children through books. You can teach them to be thoughtful and kind to animals and to love the bush." The mischief perpetrated by the Banksia Men is like the thoughtless damage that occurs when children are allowed to wander over the countryside without instruction, trampling wildflowers underfoot and catching and starving native animals.

Perhaps the one failure of *The Complete Adventures of Snugglepot and Cuddlepie* is that it does not hold the interest of native Australians. A special education teacher from the Northern Territory told me that she tried to introduce the book to her classes of aborigines. The children by and large didn't accept it and refused to conceal their boredom when she read it aloud.

Aboriginal children are perfectly willing to accept the premise that banksia cobs and eucalyptus capsules have a private life. Pure aboriginal myth maintains that all things that make up the bush have a particular personality and must be respected for it. Some aboriginal tribesmen still claim kinship with certain native animals. The characterizations in *Snugglepot and Cuddlepie,* however, are so steeped in British culture that they do not coincide with any aboriginal beliefs, so aboriginal children prefer to reject the book in its entirety.

The illustrations, however, appeal to everyone. May Gibbs's eye for artistic detail made her the envy of illustrators, who tried to follow her success, and publishers demanded that her talent be emulated. Children's author Pixie O'Harris was told that she would be published when she produced a book as good as *Snugglepot and Cuddlepie.* Her book *Pearl Pinkie and Sea Greenie* (1935) was the result. It details the adventures of rockpool sprites, who bear an undeniable resemblance to gumnut babies.

Gibbs received her highest praise after World War II when an art critic remarked, "Her careful detail (not for detail's sake, but to give that exploration for very small children who examine each part of the drawing rather than the presentation as a whole) provokes the closest, most exciting investigation. There is always something new to discover in a May Gibbs drawing." I suspect that hundreds of urban preschoolers in Australia pick up their first lessons in botany from their perusal of Gibbs's deft caricatures, despite the fact that the illustrations remain immersed in pure fantasy. You can sit on a bench along a path at the Royal Botanic Garden in Melbourne and watch children drag parents and schoolteachers from plant to plant. "Ooh, Mummy, look at the banksia!" is the usual cry, not, "What's that funny plant?" This is especially impressive when you remember that the average American child, urban or rural, can rarely discriminate a maple and an oak and that that ignorance can last a lifetime.

May Gibbs died in 1969, just short of her ninetieth birthday. She was widowed and childless, and her Sydney home, named Nutcote, was auctioned for the benefit of UNICEF as stipulated by her will. Today, profits from the sale of *The Complete Adventures of Snugglepot and Cuddlepie* help to fund the care of spastic and crippled children in New South Wales. Visitors to Nutcote find that the author was true to her own convictions. She did not attempt to transplant Australian wildflowers to her suburban garden. It was landscaped exclusively with domesticated breeds and cultivars in the English cottage style.

In some ways it's a shame she died without seeing the revolution in the breeding of native plants that has occurred in Australia over the past decade. Her books have continued to introduce new generations to the beauty of native trees and shrubs, and both amateur and federally employed horticulturists now take pains to propagate hardy forms or hybrids from seed. Almost every Australian garden can now grow a wide range of gum trees offering flowers that would be cherished in Little Ragged Blossom's wardrobe.

Chapter 3

A Country of Mistletoes

Parasite flowers illume with dewy gems
The lampless hall, and when they fade the
  sky
peeps through their winter-woof of tracery

—*Percy Bysshe Shelley,*
Epipsychidion

ONE OF THE strangest yet most haunting myths of some Australian aboriginal tribes concerns the fate of spirit babies, who are sent out into the world to find a mother. They hide in trees and rocks until women walk by. Those unable to find a mother wail dismally for a time, until they are changed into mistletoes. Legend says that the red–orange flowers are stained with fetal blood.

The myth is almost a metaphor for the natural history of the sixty-odd species of Australian mistletoe. All of them are parasites, and their fate is determined by their location and establishment upon a nurturant host. In this respect, a mistletoe seed is something like a motherless child. The mistletoes living in our own hemisphere are parasites, too, but their Australian counterparts are outstandingly successful. After five years of field study I am almost willing to admit that Australian mistletoes have elevated parasitism to a fine art.

A mistletoe is best defined as an occupation adopted by certain woody, flowering plants in the Santalales. This order also includes nonmistletoes, such as sandalwood *(Santalum)* and pin cherries or ballarts *(Exocarpos)*. There are about four families in the Santalales that contain species we would say are flaunting the mistletoe lifestyle, and two of them have representatives in Australia: the Viscaceae, or Christmas mistletoes, and the Loranthaceae, or showy mistletoes.

Viscaceae are found throughout the world, although they are rather uncommon in Australia. They produce tiny, unisexual flowers that lack petals. The European species are our cherished Yuletide decorations. The Loranthaceae are very well distributed throughout the Southern Hemisphere, however. They produce large, bisexual flowers with showy petals and are clearly the dominant mistletoe flora of Australia.

The fossil record indicates that Australian mistletoes have been a part of the vegetation there for over the last 70 million years. This suggests these parasites may have evolved prior to Australia's final separation from South America and India via continental drift. Some mistletoe groups are obviously "late arrivals or migrants" from tropical Asia that crossed the Sunda Arc, a natural landbridge whose remnants are the scattered islands of the Malaysian archipelago.

While American mistletoes are not very common throughout most of the United States, Australian mistletoes exploit almost all native trees and shrubs as hosts. Consequently, Australia is as much a country of mistletoe as it is of eucalyptus or acacias.

Mistletoes have found rather flamboyant niches in every mainland Australian ecosystem, excluding the few alpine zones. There are mistletoe species that grow only on mangroves in the tropical north, and there are mistletoes on the arid-zone shrubs that grow in the "dead heart" of the continent. On the sandy plains of the far west, one species, *Nuytsia floribunda,* found nowhere else on the planet, is a sturdy tree whose roots attack the roots of neighboring gum trees or grass tufts. So common are mistletoes that they are important food sources for some of the native marsupials that spend most of their lives in trees.

While some Australian mistletoes can live off of many hosts, others have extremely specific preferences. The gray mistletoe *(Amyema quandang)* grows only on acacias. The buloke mistletoe *(A. linophyllum)* is found exclusively on pencil pine trees *(Casuarina* sp.*). Amyema melalucae* has only been collected from the slender branches of two species of honeymyrtle *(Melaluca)* and only when those two shrubs grow along the coast.

Only a few Australian species have been reported exploiting introduced trees, but these have had an enormous impact on local silviculture. The insidious creeping mistletoe *(Muellerina eucalyptoides)* has become a familiar urbanite, harassing ornamental oaks, plums, and plane trees. Every few years hundreds of dollars must be spent to have these plants removed surgically from Melbourne's park and botanical-garden specimens.

Some Australian mistletoe species attack other mistletoes. In the Saint Arnaud region of Victoria, the succulent mistletoe *(Amyema miraculosum)* attacks only box mistletoe *(A. miquelii).* It is a most successful little plant—I have found as many as five individuals on a single box mistletoe!

Most Australian mistletoes begin life inside a specialized fruit known as a pseudoberry, which has a leathery rind surrounding a layer of sweet flesh that I think tastes a little bit like a fresh litchi. This edible layer is reinforced with a zone of sticky, indigestible threads known as the viscin tissue. Both sweet tissue and viscin encircle a single, green embryo and the embryo's stored food (endosperm). Most Australian mistletoes, like those in the United States and Europe, are unusual among seed-bearing plants because the seed lacks a true husk—the "naked embryo" is devoid of any strong, mechanical coat once the fruit ripens.

In Australia the success of the seed's unusual packaging has been attributed to the intimate coevolution of mistletoes with native birds, which has produced a remarkably efficient system of seed dispersal. Some birds gobble the ripe pseudoberries, but their gizzards grind up the delicate embryos. Other birds, such as the olive-backed oriole *(Oriolus sagitattus),* the painted honeyeater *(Grantiella picta),* and the spiny-cheeked honeyeater *(Acanthagenys rufigularis),* have far gentler innards that do little damage to a seed. I have seen spiny-cheeks after they have gorged themselves on pseudoberries. Intact seeds, cleaned of their fleshy envelopes, dangle from a bird's anus like pawnbroker's balls.

Much of the success of Australian mistletoes can be attributed to a single species of bird, the mistletoe bird *(Dicaeum hirundinaceum),* a nomadic creature that actively seeks out isolated populations of mistletoe as they come into fruit and the pseudoberries change in color from green to mealy yellow or translucent pearl. While mistletoe birds eat insects and other fruits, it seems that most of their nourishment comes from the fleshy layer of the pseudoberry. The only place in Australia lacking mistletoe birds is the island state of Tasmania, on which, not surprisingly, there are no mistletoes.

Australian naturalists have long been captivated by the pseudoberry-foraging tactics of the mistletoe bird. The bird finds a likely plant and hops about the branches, meticulously searching for ripe fruits. When it finds one, it deftly knocks off the rind the way we knock the top off a soft-boiled egg and pulls out the seed with its tweezerlike beak, viscin layer and all. The bird then swallows the seed whole, or it gums off the flesh between its mandibles. When swallowed whole, a seed is excreted within an hour, after passing *over* the bird's relaxed gizzard and moving directly into the bowel. By avoiding maceration in the gizzard, the seed is able to remain viable despite its tortuous intestinal journey.

If the mistletoe bird has pecked off just the fleshy layer, it need only wipe its beak on a twig to be rid of the viscin threads. Swallowed seeds, though, are a different matter. I have watched mistletoe birds void seeds and can almost believe they feel rather uncomfortable. A bird ruffles its feathers, squats down, and then presses its vent to its perch in a series of exaggerated abdominal thrusts while duck-walking along the branch. Because the viscin threads are smeared against the bark, seeds tend to dangle from twigs on sticky "seed threads," and since the bird's intestine is often packed with seeds, some mistletoe species are voided in little clusters of twos and threes. After watching the birds go through such contortions with the gooiest seeds (degree of stickiness appears to vary between mistletoe species), I've wondered does such a nutritious diet justify a life of violent constipation?

Viscin threads do more than act as a convenient glue for mistletoe seeds. After defecation they dry out and form a shellaclike seal over the naked embryo. At this time the seed enters a resting stage. Germination is triggered by the length of daylight cycles—different species require different light regimes. When conditions are right, the seedling pushes through its shellac shroud and anchors its "root tip," a swollen, disklike gland that releases its own natural glue, to the bark of its host. This process of sprouting and penetrating the host is still poorly understood.

Australian mistletoes are not 100 percent parasitic. Their green leaves are their own, which they grow to make their own sugars and starches photosynthetically. They are water vampires, though, who tap their hosts' water-conducting tissue (xylem), installing their own "sluice gates" and cut off the flow of water to host branch tips and leaves. In addition to taking their hosts' water, mistletoes have the added advantage of receiving the essential minerals that are dissolved in the water.

The point at which a mistletoe penetrates its host is called the haustorium. It consists of a core of water-conducting mistletoe wood surrounded by the swollen tissues of the tree host (a typical immune response). Haustoria can be either a single ball-like structure or a long, creeping limb that repeatedly penetrates its host as it grows. Cut open the haustorium and you find a beautiful ribbon of brown wood (the mistletoe portion) surrounded by the white or yellow wood of its host. Rural Australians once cut and polished these "wood roses" to expose butterfly-wing patterns, which they displayed on their mantelpieces and knickknack shelves.

It doesn't take much training to learn the difference between the leaves of an American or European mistletoe and those of its host tree. Druid priests, of course, made quite a symbolic distinction between the two and believed that the parasite was really the "soul" of the oak god. Australian mistletoe leaves are not always easy to distinguish from the host tree's, though. Some Australian species mimic their victims' leaves or photosynthetic branches to the point that suburban gardeners have vociferously objected when innocent botanists attempt to remove the parasite for study. A friend of mine was peacefully defoliating a mistletoe, because she needed leaves for a DNA study, when she was charged by an angry woman, who screamed, "Stop, stop, you're killing my tree!"

The resemblance can be remarkable and confusing. The drooping mistletoe *(Muellerina eucalyptoides)* was so named because the botanist who collected it wished to note the strong similarity between the leaf shape of the parasite and its eucalyptus host. The box mistletoe *(A. miquellii)* and the pendulous mistletoe *(A. pendulum)* produce hard, sickle-shaped leaves, which look very much like the leaves of the eucalyptus trees on which they grow. The most uncanny mimic is the buloke mistletoe, whose long, thin, grayish-green leaves are nearly indistinguishable from the long, thin, photosynthetic branches (cladodes) of its host, the pencil pine *(Casuarina).*

Since many Australian mistletoes parasitize a limited number of trees, this mimicry may be more than coincidental. Dr. Bryan Barlow of the Herbarium Australiensis and Dr. Delbert Wiens of the University of Utah suggest that mimicry is adaptive. Mistletoe leaves are food for a whole range of predators, from the caterpillars of some of Australia's prettiest butterflies to climbing, vegetarian possums. Even the camel, once introduced and now feral in Australia, grazes on these parasites. Mistletoe leaves camouflaged in their hosts' branches, Barlow and Wiens believe, run less of a risk being eaten by marsupials or located by pregnant butterflies eager to lay their eggs.

The researchers acknowledge that this masquerade may not operate over the entire range of a single mistletoe species. They suggest that it would be most adaptive in the regions where the plants evolved and where most of their predators remain. As mistletoes extended their distribution, they would encounter fewer predators, and mimicry would not be essential to survival. So, mimicry, at least for some species, may once have been more important than it is now. Additionally, over the past several hundred thousand years a number of huge, tree-browsing marsupials have become extinct. Some mistletoes may be playing hide-and-seek from animals that no longer exist.

Dr. Nick Reid, of the Department of Zoology at the University of Adelaide, theorizes that mistletoe mimicry may actually be a device to enhance seed dispersal through deceit. A mistletoe bird confronted with both the host and the parasite look-alike would have to scour the entire host tree to find all the mistletoe seeds. The longer the bird remained in the host tree, the greater would be the chance that it would defecate on a host twig. This is a valid hypothesis, because once mistletoe birds land on a mistletoe, I've noticed that they tend to wander off into the host tree as if they were searching for additional pseudoberries among the host's branches.

A third theory throws cold water on the previous two. Dr. Peter Atsatt, of the University of California at Irvine, is convinced that mimicry is just an accident of parasite physiology and should not be regarded as a form of camouflage at all. A plant transports hormones to different parts of its body through xylem, just as an animal sends hormones through its bloodstream. Trees probably manufacture most of their leaf-shaping hormones (leaf morphogens) in roots and then send them to branch tips through xylem. The water a mistletoe takes from its host is laden with leaf morphogens which is fortunate since the parasite has no hormone-making roots of its own. The mistletoe's leaves therefore grow to look like the tree's. Atsatt believes that specific concentrations of hormones are responsible for programming leaf shape, whether they are transported to the leaf buds of the tree that has made them or, second-hand, to the parasite that absorbs them. Certainly the long-flower mistletoes *(Dendropthoe)* have a wide range of host trees and have among the most variable leaf forms of the Australian Loranthaceae. They are broad and wide when they attack a honeymyrtle, sickle shaped on a eucalyptus, and long and skinny when established on a pencil pine.

Australian mistletoes are most conspicuous when they flower, which makes for a long, spectacular show even though the flowers lack a discernible perfume. Some species flower twice a year in regions that offer optimal climates and resources. Pollination can be so successful that a crop of plump fruits may be ripening while on a different branch a second batch of flowers await fertilization. Individual plants can stay in flower for months at a time. The gray mistletoe *(Amyema quandang)* has small flowers that are massed together to form striking, red-purple clusters. One plant can stay in flower for nearly half the year as it gradually but continuously replaces its spent blooms with freshly opened buds held in reserve. The harlequin mistletoe *(Lysiana exocarpi)* only displays a few fresh flowers every week, but they are large trumpets complete with a shiny scarlet "funnel" and green petal tips, which turn canary yellow as the flower ages.

Odorless blossoms and the red and orange pigments of mistletoe flowers suggest that birds pollinate them. More than sixty years ago one bright naturalist noted that Australian mistletoes were "doubly" ornithophilous; seeds dispersed by birds and flowers pollinated by birds. In our hemisphere, the Christmas mistletoes depend on a combination of wind and insects to carry their pollen. In contrast, the Australian mistletoes, evolved in the absence of social hive bees, have exploited a feathered resource hungry for nectar. Each mistletoe flower in Australia contains a nectar gland that forms a tiny crown or collar around the neck of the pistil on the floral base. Representatives from five bird families land on mistletoe twigs and force their bills down the floral throats in search of nectar secretions.

In fact, so many different kinds of songbirds consume nectar as a primary or secondary source of energy that mistletoe pistils rarely suffer from neglect.

Birds in the family Meliphagidae (honeyeaters) are probably the dominant pollinators of mistletoe flowers in Australia. Honey-eaters are true perching birds, or passerines, more closely related to wrens or crows than to hummingbirds. There are more than seventy-five honeyeater species in Australia, ranging from wren-sized birds to shrill, bossy fowls often larger than a domesticated pigeon. The triangular or trilobed pollen grains of the mistletoes either adhere to their probing bills or catch in the microscopic feather barbules of their face and breast plumage. Because honey-eaters have sharp needlelike bills and tend to gather nectar in quick, jabbing motions, mistletoe blossoms have tough, hard pistil tips and petals reinforced with thick, waxy coats.

Colonies of flowering mistletoes are excellent places to bird-watch from dawn to midday. Not only can you appreciate honey-eater size, color, and diversity, but there are usually foragers from other families, such as flocks of delicate silvereyes (Zosteropidae) and parrots like the flower-munching rosellas *(Platycercus)* or the less destructive lorikeets (Loriidae). Mistletoe flowers tend to be so accessible to nectar drinkers that it's not unusual for more than a dozen species of birds to pollinate one species of mistletoe. Of course, a dozen different bird species do not turn up at the same mistletoe at the same time. Flowering periods are often so long that they coincide with the migratory routes or nomadic wanderings of different honeyeaters. Spinebills *(Acanthorhynchus)* are among the first visitors to pendulous mistletoes *(A. pendulum)* during the last

Flowering branches of a mistletoe from tropical Queensland, Australia. The petals form a long, interlocking tube with nectar secreted at its base. Petal tips pull backward, exposing a bright orange surface, indicating that the odorless flowers are ready to receive birds with long, probing bills. Photograph by Trevor Hawkeswood.

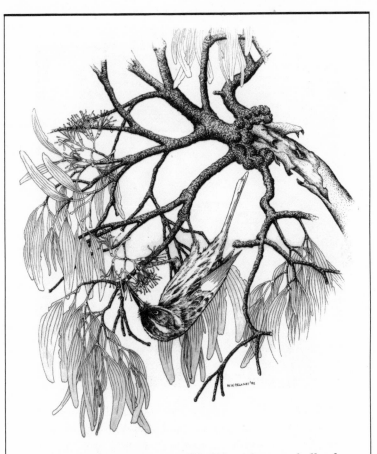

A New Holland honeyeater (Phylidonyris novaeholland-iae) *takes nectar from the flowers of the box mistletoe* (Amyema miquelii). *The bird grasps the branch with its claws, hangs upside down, and probes the floral tubes with its bill. The sexual organs of each flower brush the bird's bill and face plumage. Notice the "ball" haustorium the mistletoe has formed on its gum tree host.*
Illustration by W. W. Delaney.

There are different ways in which the pollen grains of mistletoe flowers can become attached to their bird pollinators. Each grain wears a single, sticky glob of pollenkitt (a fatty cement), which may attach the grain to feathers or beak. However, since each grain has three rounded arms, an arm can also be caught between the feather barbules of the face and breast plumage. Photograph by P. Bernhardt.

winter month of August, but they may be replaced by yellow-faced honeyeaters *(Lichenostomous chrysops)* in early October.

Avian aggression also limits the number of birds allowed to forage on a parasite. When nectar production peaks, some honeyeater species protect their favorite food sources by chasing away competitors. Such "turf wars" are acrobatic, colorful, and surprisingly unpredictable. I couldn't always pick the winners, as there were few fair fights. It's true that the bigger species of honeyeaters chase away the little guys, but family flocks of smaller birds, such as the fuscous honeyeater *(Lichenostomous fuscus),* often gather enough courage to mob the bullies and drive them away from the mistletoe colonies.

The life cycles of mistletoes have allowed them to enjoy spectacular success in Australia, whose ancient soils have been leached of large quantities of nutrient salts by millions of years of erosion and climatic change. While Australian trees and shrubs, such as the gums and banksias I discussed in the last chapter, have evolved complex systems to recycle nutrients, mistletoes and other parasitic plants simply rob their hosts. This suggests quite a different evolutionary rule of thumb: "If you can't find your own essential resources, steal them from someone who has." Since mistletoes exploit their hosts as living pipelines, they have never needed to develop roots that could survive in cement-hard soils. Over the last 70 million years, as the continent has gradually dried out, this attribute has become increasingly adaptive. There are other families of parasitic plants in Australia, but none show the wide distribution, diversity, or density of the Loranthaceae.

Successful as the mistletoes have been on their own, recently they have enjoyed a population explosion due to tree thinning for timber or agricultural purposes. The great gaps created in the forest canopy have boosted mistletoe growth and stimulated their levels of reproduction. With the increased light through the canopy, mistletoes do not have to compete with their shady hosts for sunlight. In addition, converting woodlands into savannas translates into fewer trees, which means fewer perches for mistletoe birds, so the surviving trees often suffer from higher rates of infection.

Infestations can reach a point where more than 60 percent of a tree's canopy is not tree at all, but mistletoe. One or two mistletoes on a mature gum tree will probably not do much damage, but when a tree's roots support more mistletoe leaves than tree leaves, the results can be fatal. Mistletoes weaken their hosts, encouraging the spread of serious diseases or lowering the tree's resistance until branches succumb to wood-boring beetles. Mistletoe infestations have probably contributed to the dieback of famous old stands of trees in Australia.

The fact that people just don't recognize mistletoes may be another reason for mistletoe spread. There were times when I told people in Melbourne I was studying mistletoes: "Fascinating," was the usual reply. "Does Australia have any native kinds?"

Oddly enough, while North American mistletoes are rampant pests of the western pines and juniper groves, cone-bearing evergreens remain the one group that Australia's mistletoes seem unable to dominate. In this dry continent where Christmas comes in summer, the peaceful coexistence between Australian mistletoes and imported pine plantations seems like a plant version of the Peaceable Kingdom.

Chapter 4

Prairie Days

I loafe and invite my Soul,
I lean and loafe at my ease,
observing a spear of
summer grass.

—*Walt Whitman*

"LAST EVOLVED, FIRST dissolved," is a cynical but concise way of describing the history of our native grasslands. We condemn those who exploit and destroy the exotic rain forests of South America, but the history of our own sodbusters remains cloaked in reverential sentimentality. These pioneers successfully transformed a huge chunk of this country into a national breadbasket, but at the loss of driving the North American prairies to the brink of extinction.

Just a minute, I don't want to spend a whole chapter composing a condescending elegy for a bruised biome. This is a botanical romance, and I want to introduce you to the most recent object of my affections. She lives in the Flint Hills of Kansas, all 8,616 acres of her, and her proper name is the Konza Prairie Research Natural Area. Konza is managed by Kansas State University, and the original purchase was made possible by the philanthropy of Ms. Katharine Ordway back in 1977. Yes, this is the same Katharine Ordway known for her unstinting patronage of the artist Mark Rothko.

Last spring I spent two weeks on Konza and was seduced by the ecology of the tallgrass prairie. This plant community may just be a remnant of its former glory, but it is a unique remnant. A prairie is not a neglected field of lawn grass that needs mowing. You can't get one from planting one of those cans of seeds available in nursery shops. Grasslands are complex, extensive, self-maintaining environments whose plants employ some of the same adaptive tricks I've discussed in the preceding chapters.

Our American prairies are babies of the Ice Age. The plant communities of Konza, for example, must be less than 12,000 years old. Evidence of layers of fossil pollen in the Konza soils shows that as the last of the glaciers retreated from Kansas, climatic shifts caused steady but radical changes in the vegetation of the Flint Hills. Less than 11,400 years ago, aboriginal, spruce/pine forests had already been replaced by oak woodlands, and the oaks, in turn, had fallen to the advance of true grasslands. Konza's 600-year transition from snow forest to scrubland to true prairie is regarded as rather sluggish by some geologists, who surmise that adjacent regions converted to grass in less than 200 years following the glacial retreat.

Due to the interrelated extremes of geological and climatic change during the Pleistocene, the grasslands inherited the deepest and richest topsoils in this country. Such soils are perfect for intensive agriculture, but this has meant the ultimate destruction of most of the prairie habitats. Konza escaped much of the agrarian cataclysm because the limestone ridges of the Flint Hills wear a much shallower crust of topsoil. The farmers on this land were pastoralists, who only suffered the plant communities to become food for their cattle. The land may have been overgrazed, but the grasslands were not torn up for cereal crops.

Grasslands then once straddled both sides of the Mississippi River, forming a wide, natural barrier between the deserts of the Southwest and the deciduous forests of the East. Because it receives up to 40 inches of rain a year, a true grassland is too moist to encourage the spread of arid-zone shrubs and succulents, yet it is still too dry to support stands of taller, temperate hardwoods. Within the American grasslands are two distinct but overlapping communities. The western range, which receives less rain than the East, is known as the Great Plains or steppes and is clothed in the shorter species of bunch or buffalo grasses. The East receives a more generous rain shadow, and Konza remains rich in the species that comprise the tallgrass prairie. At one time tallgrass communities extended from Illinois north to Minnesota and then west from Missouri to the edges of the Dakotas, Nebraska, and Oklahoma.

How tall are the tallgrasses? Well, this question can't be answered properly without just a little knowledge of the anatomy peculiar to members of the grass family (Poaceae). The tallgrass species of Konza are a bit like icebergs, since most of their mass remains below the soil surface. You can't dig very far without hitting the thick, jointed, underground stems (rhizomes) of each tallgrass species. These rhizomes live year after year and grow horizontally through the ground. Eventually different clumps collide and intertwine. Roots emerge from the rhizome joints and penetrate the soil, forming a fibrous net. Each spring the rhizomes of the tallgrass species produce erect, secondary shoots called tillers. Tillers push through the soil and grow upward to become the tall, flowering stems, or culms. In a fair year the culms of the tallgrasses will grow chest high before they bloom. Indian grass *(Sorgastrum nutans)* can grow to 6 feet. Big bluestem *(Andropogon gerardii)* can put Indian grass to shame, as its culms can stand 8 to 10 feet high at Konza when rains are generous.

Culms live only a single season no matter what the species of tallgrass. During the summer, food made in the culm leaves is transported down to the starchy rhizomes for winter storage, so that the dry, dead stems remaining above ground in autumn have no more nutrients than natural straw. That is why great herds of American buffalo once migrated west in winter. The short bunchgrasses of the western plains lack storage rhizomes, so nutrients collect in the short tufts of dormant shoots. Such shoots afforded the great herds a plentiful source of winter hay.

On the tallgrass prairie, winter rain and snow beat down the dead culms until they form a thatch over the ground. Thatch buildup over the years can cause changes in the plant community. Layers of thatch act as sun reflectors, so it takes longer for the soil to warm up each spring, and this delays rhizome activity. The layer can also prevent water from soaking into the soil, slowing the development of new tillers. It's ironic that tallgrasses can actually suppress their own growth because of thatch. If the layer is left, the prairie can be colonized by invasive bunchgrasses and small, scrubby bushes, such as the red cedars *(Juniperus virginiana)*.

Cyclical fire shapes tallgrass communities in much the same way it influences the proportion of trees to shrubs in Australian woodlands. Konza has been subdivided into many test plots for plant-diversity studies between burns. Since the tallgrasses survive as underground rhizomes, they are not damaged by fires in early March or April. Neither are the dormant bulbs, tubers, and taproots of the many wildflowers that will grow among the grasses. Fire removes the dry mulch of thatch and kills the exposed shoots of bunchgrasses while turning resinous branches of cedar into living torches.

Since the tallgrass prairies are where the deer and the antelope play, one would expect that grazing and browsing by mammals in summer would reduce the trend toward thatch buildup. This isn't true, though, not even on Konza, where buffalo herds have been reintroduced. Ecologists estimate that both the large, hooved mammals and the many rodents and rabbits eat only about 5 percent of the summer's green growth. The real limits to plant growth, it turns out, are foraging insects and uncountable numbers of roundworms, or nematodes, which live in the soil and burrow into plant roots and rhizomes to suck their juices.

The plants of the tallgrass prairie struggle silently to reduce the damage inflicted by so many marauders. Grasses incorporate particles of silica into their cell walls. This grit probably wears down the mouthparts of diners. A few wildflowers, such as the prairie rose *(Rosa setigera),* bull thistle *(Cirsium undulatum),* and sensitive brier *(Schrankia nutallii),* either arm their tough stems with thorns or line their leaves with irritant prickles or spiny tips. Chemical defenses may rival the trees of the lowland forests of Central America. The flowering heads of compass plant *(Silphium lacinatum)* and rosin weed *(Grindelia squarrosa)* are protected, in part, by sticky, ill-smelling secretions, nasty surprises released once they are bitten.

Different species acquire poisons in different ways. Most appear to manufacture their own secondary compounds. The poison suckleyana *(Suckleya suckleyana),* a member of the goosefoot family (Chenopodiaceae), offers leaves rich in cyanide. The roots of other species trap metal ions, which are then absorbed into the plant body until they reach toxic concentrations. Fields of poison vetch *(Astragalus pectinatus)* may accumulate enough selenium to kill sheep or goats.

In fact, European colonists soon learned that innocent-looking patches of wildflowers could weaken or kill their stock. Consequently they gave some of the prettiest blossoms common names that would have delighted the Emperor Nero. Many of the native members of the sweet-pea family (Fabaceae) are known as loco-weeds. Death camass *(Zigadenus nuttallii)*, snakeroot *(Eupatorium rugosum)*, and stinking milk vetch *(Astragalus praelongus)* are just a few of the names that convey both a warning and a terrible sense of exasperation.

It's easier to understand why plants evolve increasingly viru-lent poisons once you've observed the levels of damage inflicted by plant predators. I spent two weeks studying the reproductive biol-ogy of the death camass. About 30 percent of the creamy white, lilylike flowers on each plant were fertilized and began to develop capsules. This is a most respectable rate of fruitset in herbs cross-pollinated by small bees. However, only about 1 percent of those capsules grew large enough to release their seeds. The remaining ovaries were devoured by the yellow and black caterpillars of a geometrid moth. To add insult to injury, these larvae hung threads of dung-encrusted, filthy, gray silk all over the fruiting stalks.

Do you have the impression that the plant life of the tallgrass prairie may be rather dynamic? The rise of the tallgrass prairie appears to have been of major benefit to the distribution of some of the major genera of American wildflowers. Prairies acted like magnets to herbs that had to compete with trees or spreading shrubs on other parts of this continent. Species that would be confined normally to a few individuals in forest glades or temporary meadows have established extravagant colonies and clones on true grassland.

When you walk along the Konza paths, you notice that the tallgrass prairie is really a series of interlocking communities, and each plant species tends to have a rather patchy distribution be-cause of discrete differences between local habitats. The abundance of wildflowers varies according to such factors as drainage, expo-sure to the sun, and the interplay of thatch versus cyclical burns. For example, Ted Barkley, the current director of Konza, knew that I wanted to see the diminutive nipple cactus *(Coryphantha missouri-ensis)*. My wife and I followed him for an exhausting mile up and down the Flint Hills. Nipple cactus grows along the dry, limestone ridges where the tallgrass doesn't thrive. It was worth the extra sweat to find the cacti in full bloom with their vividly violet stamens set against a yellowish-bronze background of pointed petals.

Large animals may play a role in the maintenance of prairie flowerbeds. Ted introduced us to the new herd of buffalo—forget what you've seen in John Wayne movies. These bison don't trample everything in their paths. Ted showed us that while the animals have thick, wide bodies, those bodies are mounted on rather slender legs and comparatively narrow hooves. We were tickled to see how they stepped delicately through the vegetation without seriously crushing shrubs or thick patches of herbs. As landscape architects, herd members make dust or mud wallows and will dig up the ground to establish salt licks, disrupting standing vegetation and allowing seeding of exposed sites once the buffalo abandon them. Wallows are colonized temporarily by a number of low-growing plants before the tallgrasses reinvade. Sedge species *(Carex)* often occur most abundantly on old wallows. Over 130 sedge species are distributed throughout the American grasslands.

The color and form of the tallgrass prairie depends on both the time of day and the time of year. When I left the field station, an hour or so after dawn, the purple pom-poms of the sensitive briars had just opened for the day. The burgundy cups of the poppy mallow *(Callirhoe papaver)* did not expand until midmorning or early afternoon. Sundown was my favorite time, as I could watch the unfolding of the fluttermills *(Oenothera macrocarpa)*. This species of evening primrose appears to be pollinated by sphinx moths, but each large flower rarely lives more than one night. Buds open dramatically at dusk—greenish sepals curl back exposing the pale gold of the petals, which are fused together to form a trumpet. The mouth of the trumpet unfolds with the slow, synchronous, and clockwise spin of the petal lobes.

Wildflowers grow as tall as neighboring grasses because to attract pollinators, most flowers cannot afford to be hidden beneath grass stems and leaves. Floral stalks tend to be just a little bit taller than the tallest tiller or culm. In late March or early April, most new tillers are still waiting to poke above the soil. Consequently, low plants, such as the pasqueflower *(Anemone patens)* and the prairie plum *(Astragalus crassicarpus),* can blossom at ground level. By the end of May the tillers of tallgrass reached just above my ankle. The milkweeds *(Asclepias),* lead plants *(Amorpha),* and death camass were halfway to my knee, though. By the time the culms of Indian grass and big bluestem set seed in late summer, the tallgrass prairie is colored by the massive flowering heads of the stalkiest members of the aster and sunflower family (Asteraceae). Eleven out of the thirteen prairie species of sunflower *(Helianthus)* are distributed throughout Kansas. (See color plates 3 and 4.)

Grasses flower, too, and this surprises a number of people. Most of us have never seen them do this. The paired flowers spend most of their brief lives hidden behind overlapping bracts, in units known as spikelets. If you rise early on a dry, sunny morning, you can watch the spikelets open as the inner bract expands and the dangling stamens and feathery tips of the pistils peep out from between the outermost bracts, or glumes. Some scientists claim they've heard a little crackling noise as hundreds of grass flowers emerge simultaneously. Each blossom usually consists of just three stamens and one pistil, a very simple design. There are no colored or perfumed petals or sepals, since almost all members of the grass family shed their pollen on even the mildest of air currents.

If you want to appreciate the frugal beauty of grass flowers, you must always bring a hand lens. The grass pistil usually bears two receptive tips, both of which have luxuriantly plumed, feather-boa-like stigmas. These twin structures "sift" the air for the whirling grains of pollen released by chimelike stamens, which tip out their flattened grains within the first few minutes of blossoming. Most grass flowers are allotted no more than a quarter of an hour to complete pollination before the glumes shut them up for good. The dry, single-seeded fruit that develops behind the glumes is known as a caryopsis. Each grain of wheat and kernel of corn is also a caryopsis, as all are members of the Poaceae.

The pollen grains wear a thin protein film, which they use as an identification card when they land on the appropriate stigma. Without the correct protein composition the stigma will not allow the grain to germinate and produce its pollen tube so that the sperm can travel to the ovary. Unfortunately, when we inhale too much pollen, the protein film can cause an allergic response. Residual grass pollen, trapped in the atmosphere by winds, remains a major cause of hay fever.

I'm willing to live with the prospect of a runny nose if I can return to Konza every year. The reserve is not open to the general public, so all who enter must be prepared to do serious research if they want to stay. There are worse fates than laboring among the tallgrasses, and I look forward to a new selection of botanical mysteries when I visit again.

— Part Two —

The Floral Theater

Chapter 5

The Forms of Flowers

De sunflower ain't de daisy
And de melon ain't de rose
Why is dey all so crazy
To be sumpin' else dat
grows?

—*Edwin Milton Royle*

ONE OF MY first responsibilities as a novice botanist was to take tour groups over the grounds of the New York Botanical Garden and through the Enid Haupt Conservatory in the Bronx. Visitors were treated to running commentaries on flower folklore, life cycles, horticulture, and a smattering of plant classification. Not everyone appreciated my attempt to introduce them to the potted version of more than two centuries of nomenclature and taxonomy. In fact, one visitor was rather hostile.

It happened on a fine spring day as I was showing her group some of the early flowering shrubs and bulbs in the rock garden. Some, I pointed out, were members of the magnolia family (Magnoliaceae), whereas others belonged to the family of wake robins and tulips (Liliaceae).

"How can you just keep saying that this plant belongs to that family and that plant belongs to a different family!" an angry voice in the crowd snapped.

I should have known that her remark was not really meant to be a question, but I hadn't been on the job long, and I offered what I thought was a reasonable explanation. "The same way you can say that a house cat, tiger, and lion belong to the mammal family called the Felidae," I replied. "Different plant species, like animals, share unique anatomical features that suggest a common origin, and we acknowledge their similarities by clustering them within the same family. We put a house cat, tiger, and lion in one family because they share similar skeletons, eye structure, pelts, and other features. But we can exclude other flesh-eating, four-legged mammals such as dogs or raccoons."

The woman ignored me and chatted with her friends for the rest of the tour. I later learned she was the hostess of a gardening show on Long Island radio.

If most people have difficulty recognizing natural alliances within the plant kingdom, I can sympathize with their confusion and impatience. Plant classification seems rather arcane, even though botanists tend to use much the same sort of tools as zoologists. Field guides to the wildflowers, however, come equipped with pitfalls absent in animal field guides. Body shape is usually a dependable tool in the recognition of larger creatures with backbones. Bird-watchers become so adept at identification that they can recognize species by silhouettes as they fly in front of a full moon. Flower form and floral presentation, on the other hand, play tricks on the best of memories. De melon ain't de rose, but it would be easy to confuse a rose with quite a different species in another family even if plant breeders didn't add more petals, change colors, and delete extra sexual organs to make blossoms of different origins look alike.

At first glance a rose and a camellia flower appear quite similar to each other. They even share much the same range of color. They are not closely related, though. The rose is in the family Rosaceae and has far more in common with the blossoms of raspberry *(Rubus)* or strawberry *(Fragaria)*. The camellia is part of the family Theaceae and is a closer relative of the shrubs that give us commercial tealeaves instead of tea roses.

Why do unrelated plants bear flowers with such similar shapes? This question was addressed in a pioneering work of the 1960s, *The Principles of Pollination Ecology.* Its authors, Knut Faegri and Leendert van der Pijl, set forth what is now a well-accepted concept: plants offer a limited number of floral advertising forms. Briefly, certain flower forms attract one or more kinds of animal pollinators. Adopting one of these forms assures efficient pollination for a plant species, because insects, birds, and small mammals learn that returning to similar flower shapes will provide a reliable source of food.

In setting out the concept, Faegri and van der Pijl described six basic floral forms that are used over and over again by many different plant families. Form is determined by the size and shape of the outermost ring of sepals (the calyx) and the inner ring of petals (the corolla). Collectively, these two rings of floral segments are known as the perianth.

The oldest and simplest of all floral advertising forms is known

Salver/bowl form expressed in a woodland anemone (Ane-mone) (top). Bell/funnel is expressed in two ways. The nodding bell is a bellwort (Uvularia), while the angled blossom is from a European columbine (Aquilegia). Each of the five petals forms a separate funnel with its own hooked tip (bottom left and right).
Illustrations by J. Meyers.

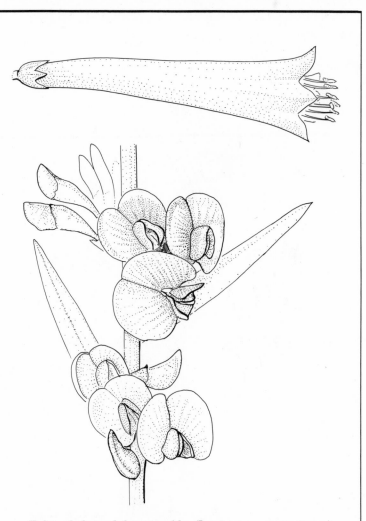

Tube of the red honeysuckle (Lonicera sempervivans) *lacks a landing platform. The nectar is concealed at the base of the floral cylinder (top). Flag flowers of a bitter pea* (Daviesia). *One petal develops into the large, erect "billboard" known as the banner (bottom).*
Illustrations by J. Meyers.

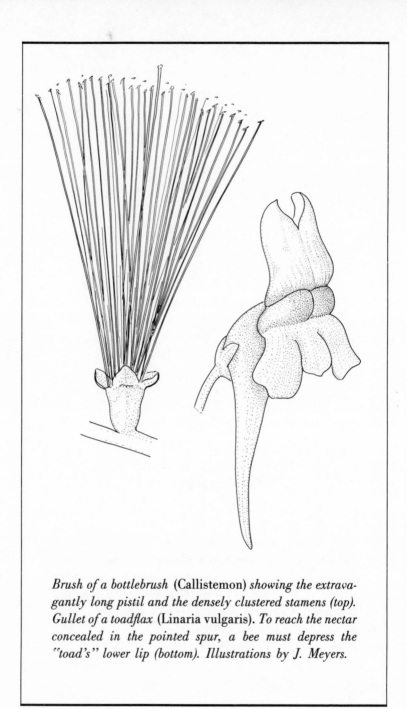

Brush of a bottlebrush (Callistemon) *showing the extravagantly long pistil and the densely clustered stamens (top). Gullet of a toadflax* (Linaria vulgaris). *To reach the nectar concealed in the pointed spur, a bee must depress the "toad's" lower lip (bottom). Illustrations by J. Meyers.*

as the salver/bowl. As the name suggests, the perianth segments are arranged in a pattern resembling a serving dish or a soup bowl. The flower is held erect on its stalk (pedicel). Obvious examples of a salver/bowl form include the wild roses or eglantines *(Rosa)*, cinquefoils *(Potentilla)*, camellias, and poppies *(Papaver;* see chapter 7).

The second form is the bell/funnel. The overall form is much like that of the salver/bowl, but the perianth organs tend to be longer, united, and may or may not stand stiffly at attention. Bellflowers nod on their stalks with the open end facing downward. Their perianths tend to be rounded or concave at their bases. Typical bellflowers include the fuchsias *(Fuchsia)*, mountain laurels *(Kalmia)*, and lily-of-the-valley *(Convallaria)*. Funnel blossoms may be presented as erect, nodding, or horizontal. Unlike the bellflowers, funnel perianths are constricted at their bases. Furthermore, some funnel flowers have a wide, curving outer rim. The blossoms of wake-robins *(Trillium)*, morning glories *(Convolvulus)*, and tobacco *(Nicotiana)* are among the many obvious funnel forms.

The third floral advertising form is the tube. Tubes lack the tapered rim of the funnel flower, and the perianth of the tube forms an exaggerated cylinder or canal. Tube flowers may be very long, like those of the pineapple family (Bromeliaceae) and scarlet honeysuckle *(Lonicera sempervivens)*. Tiny tubular flowers are often massed together to make massive flowering "heads," like those of the sunflowers *(Helianthus)* and other members of the daisy family (Asteraceae).

The fourth form, the brush, has a perianth conspicuous by its total or near absence. Perianth segments fail to develop, or persist only as scalelike vestiges. What we see are flowers made up of many long, delicate sexual stamens and/or pistil necks (styles). Brush flowers are often massed together in dense pom-poms, fuzzy rods, or candelabras to the pollinators. Pussy willow *(Salix)*, sensitive briar *(Schrankia)*, and century plant *(Agave)* produce brushes.

Each of the four previous forms show radial symmetry. In contrast, the flag form has one or more petals that are longer and wider than others in the same ring. This results in a blossom that has irregular symmetry. Most of the members of the pea and bean family (Fabaceae), which belong to the subfamily Papilionoideae and include alfalfa *(Lotis)* and bluebonnet *(Lupinus)*, have a wide, specialized "banner petal," which forms the flag. Similar flags are produced by some of the milkworts *(Polygala)* and water hyacinths *(Eichornia)*.

The final advertising form is the gullet, and it shows complete bilateral symmetry. The flower has a discernible front and back, like the head of an animal. It begins with a pair of pursed "lips" and follows with an expanded "throat" that often tapers off into a pointed, hollow spur. Sometimes the perianth segments form both pairs of lips, as in many of the members of the snapdragon family (Scrophulariaceae) and in a number of mints, such as the *Salvia* species. In other cases sexual organs combine with each other to form one lip, and they lay on top of the second petal lip. This is the dominant form of floral presentation in the orchid family.

Flowers do exist that span the categories, blending the characteristics of two or more of the original forms. For example, a wild iris has a flower in which each of the three broad lobes of the pistil (stigmas) form three upper lips, which lay on top of three equally broad sepals. The flower at first glance appears to be a three-headed gullet. However, the three sepals are much wider and more brilliantly colored than each of the pistil lobes. Therefore, an attractive flag dangles down from the lower lip of each gullet.

Some plant families exploit only one form. The water lily family (Nymphaeaceae) depends almost exclusively on the salver/bowl. Other families show extraordinary variation in the number of floral advertising forms they express. The buttercup family (Ranunculaceae) may express as many as five out of six of the forms. The yellow kingcup *(Caltha)* is a salver/bowl flower, whereas the nodding columbine *(Aquilegia)* exhibits the bell/funnel form. Meadow rue *(Thalictrum)* forms branches of dense brushes, and the enlarged helmet of the monkshood *(Aconitum)* incorporates at least part of the flag form. Our native larkspurs *(Delphinium)* show all the traits of a gullet. All this depends on how floral development is controlled during the earliest phases when floral

Bell/funnel-form flowers discourage less-specialized animals. Nectar glands are usually concealed at the base of each blossom. In order to reach the sweet secretions, a pollinator must be capable of hanging upside down on a bellflower. Even bats that pollinate the tropical bells must cling to the nodding perianths with sharp claws on each wing "thumb" and with their toenails. Insects that are successful foragers on erect funnel flowers use the funnel's widened rim as a landing platform, but they still must have tongues long enough to reach the nectar glands at the base. This is how large butterflies negotiate the funnels of candytuft *(Phlox)* and lantana *(Lantana)*. Horizontal funnels, such as trumpets, don't always offer a significant landing platform, since the rim may not be wide or steady enough to offer the forager a perch. Successful pollinators of such floral funnels as those of the night-blooming orchids (see chapter 16) usually extract nectar on the wing, never actually landing on the rim.

In many ways, then, the large tube flower is the ultimate in exclusivity. It lacks a landing rim, and the long, narrow canal must be navigated by a long, thin tongue. Pollination of such flowers has become the province of hummingbirds and hawkmoths, as well as the tangle-vein flies *(Nemestrinidae)* found in Africa. These animals have long, probing bills or tongues—sometimes longer than the animal's body—and they can feed while hovering in midair.

The tiny tubes that make up drifts of goldenrod *(Solidago)* or asters *(Aster)* are an important exception to the rule that blossom size and pollinator size must correspond. The minuscule tubes are so densely massed that the effect is similar to that of one very large flower. Each flower contains an exceptionally small quantity of nectar at the base of its short canal, but, taken all together, they are a feast for small flies and bees as well as large wasps and monarch butterflies, which walk across the compound platform of unified tubes casually probing for nectar.

In the case of brush flowers, the pollinator is dusted or lathered with pollen. These flowers do not offer a flat landing space, so the animal must either clutch the tough sexual organs like ladder rungs or hover before the living powder puff. Animal-pollinated brush flowers often have stamens and pistils that are brightly colored and offer either copious supplies of pollen to eat or dripping nectaries. Without petals or sepals wide enough to wear fragrance glands, a number of brush flowers appear to release their aromatic oils from the wrinkled tips of the anthers. Not all brush flowers are pollinated by animals, though.

The majority of North American trees with brush flowers, such as oaks *(Quercus)*, birches *(Betula)*, elms *(Ulmus)*, and aspens *(Populus)*, depend completely on air currents to accomplish their pollination. Such trees blossom in early spring because pollination and fertilization are more likely to succeed before the leaves unfold. Leaves create a natural "windbreak," which impedes the movement of airborne pollen. These Temperate Zone trees are not the only plants to offer brush flowers anticipating gentle gusts, however. The grasses in chapter 4, as well as marijuana *(Cannabis)*, sedges *(Carex)*, and a number of spurges *(Croton)* show similar modifications. These brush flowers are much reduced in size, are dull green or straw white in color, and they are often unisexual, with male and female flowers segregated on different branches. The pollen is adapted for shooting the breeze, not riding the legs, hairs, or beaks of animals.

It is this buoyant pollen of unobtrusive brush flowers that is the real culprit of severe, hay-fever-type allergies. Every autumn benign fields of goldenrods and asters are accused of aggravating sensitive lungs and mucous membranes and then sprayed or mowed down. The only real "crime" of these insect-pollinated tubes is that they have chosen to flower at the same time as the unassuming but noxious ragweeds *(Ambrosia)*.

Pollination in the flag-form blossom is accomplished with the help of nectar guides, flag petals decorated with contrasting colors, dots, and lines, which indicate the presence and position of nectaries. Actually a better name for this advertising form would be billboard flower, since the largest petal acts like a poster to attract passing bees (the commonest pollinators of this form). In some plants the flag petal is also a landing platform, and in certain orchids the flag is cupped and reinforced to better support the bee.

The pollination of a gullet flower often demands a determined effort, even a feat of strength from the pollinator. The gullet lips in snapdragons and in many mints must often be forced open. As if this weren't enough to discourage all but the most rugged of visitors, the gullet nectaries are lodged deeply, at the base of the throat or the tip of a spur. The pollinator must have a long tongue as well as a strong, heavy body. Bees easily seen with the naked eye like bumblebees, honeybees, and the noisy carpenter bees tend to be the most effective pollinators of the gullet advertising form.

One of the adaptive advantages of the repetition of floral forms is that plants growing within the same habitat can exploit the same pollinators but avoid competition by staggering their respective flowering periods. In North America the gullet forms of our larkspurs *(Delphinium)* yield their bumblebees to the gullets of toadflax *(Linaria)* by midsummer and the purple gullets of loosestrife *(Lythrum)* by late summer or early autumn. Conversely the mixture of different floral forms displayed by different species during the same season indicates, to some extent, that communities composed of different species have divided up the finite number of pollinator species living within a shared habitat and season.

Similar situations prevail in the tropics, where both the growing season and the number of plant species are greatly augmented. Some tropical regions always seem to be in flower, providing a steady food source for warm-blooded animals such as hummingbirds that depend on nectar for calories. To feed the over 160 species of hummingbird found along the equatorial belt, the forests offer a bewildering annual array of scarlet tubes, bells, and funnels. Even the cactus, bean, and orchid families, so commonly associated with bee pollination in our own hemisphere, have evolved species that accommodate hummingbird bills in the New World tropics. This often involves altering ancestral bowl, flag, and gullet forms: bowls can be elongated until they form narrow throats; the banner of a flag can be reduced in the bud until it is little more than a sheath around a tube; prop open the lips of a gullet, "puff out its cheeks," and you have a convincing funnel. Someday, perhaps, a talented mathematician will have a lot of fun analyzing the topology of floral forms.

How closely can convergent evolution of floral forms make unrelated species resemble one another? Parallelisms can puzzle even professional botanists. For over a century most plant taxonomists placed several tree families, such as the willows (Salicaceae), walnuts (Juglandaceae), and birches (Betulaceae) within a larger unit known as the Amentiferae. All of these trees shared the same petal–absent, wind–pollinated brush flowers, with tufted, unisexual spikes called catkins, or aments. These twig-bearing catkins suggested a very natural alliance to German botanists like Adolph Engler. Engler went as far as to develop a system of classification in which almost any tree or shrub with small flowers that lacked petals could be derived ultimately from hypothetical ancestors with catkins. Only within the last twenty or thirty years have comparative studies of wood, bud development, and fruit morphology by plant anatomists and morphologists exposed the Amentiferae for the unsystematic "attic" it really is.

Far from being the independent food-making factories that we think they are, flowering plants demonstrate a subtle dependence on both animals and inanimate forces that control critical aspects of their reproduction. One aspect of this ancient relationship manifests itself in the repetition of floral forms throughout the green world. Avid gardeners and flower arrangers insist that each flower is unique. Allergy sufferers and the florally apathetic remain unimpressed by blossom variety. Both attitudes represent extremes of taste, since "sculpturing" by natural selection has produced a bouquet in which variation sets its own limits.

Chapter 6

Pollinating Possums

The little mouse, therefore, is "the stealer."
Pliny the Elder recommended mouse ashes
mixed with honey to soothe an earache.

—*Mary Durant*

THERE ARE A limited number of ways of reaching flowers that hang from slender twigs or terminate a tall stem. The last chapter suggested that wings are an essential piece of equipment for any animal involved in the intimate act of floral pollination. Even after the sun sets, the beat of wings goes on as hawkmoths, settling moths, and bats flock to nocturnal sources of nectar.

Observations made in the shrublands and woods of Australia indicate that this continent is the home of several special mammals that are important exceptions to the wing rule. Field and laboratory research, completed within the last decade, shows that several species of marsupials pollinate flowers without recourse to aerodynamics. These relationships are not only interesting for the sake of their sheer novelty, but they may also be descendants of some of the earliest trends in the evolution of flower-animal interactions. Plant pollination by mammals is known collectively as therophily. The study techniques and theories pioneered in Australia have helped to uncover startling examples of therophily in several other regions of the world.

The honey possum *(Tarsipes rostratus)* is a marsupial found only in the southwestern corner of the Australian continent. In Australia almost all climbing marsupials are properly called possums, never opossums, which is a North American word for its single native marsupial. We know more about this member of the family Tarsipedidae than any other wingless pollinators. It's a tiny beast that looks like a long-nosed mouse but is a true marsupial that spends three months of its life in its mother's pouch.

Honey possums spend most of their lives in the trees and shrubs. They are equipped with four grasping paws and a prehensile tail. Close inspection of the tubular mouth shows just how little this animal resembles a true rodent. Force open the long snout and you'll be surprised to find no teeth at all, except for two "stubs" located at the front of the ridged palate. "A cornered honey possum," says Western Australian ecologist Ron Wooler, "tries to gum you to death." Honey possums do have a long, keeled tongue, tipped on its upper surface with many bristlelike cells.

When kept in captivity, honey possums will eat a few soft-bodied insects. Field studies show, though, that pollen and nectar form the complete diet of adults in the wild. Honey possums climb the branches of favorite food sources to lick nectar from flowers with their absorbent brush tongues. Pollen adheres to their fur and, because they vigorously preen themselves about once every fifteen minutes, is frugally licked up and devoured. The grains caught in the bristle cells of the tongue are dislodged when the possum rubs his tongue against the frontal stubs of the palate. Pollen is the honey possum's only source of protein and fat.

The search for nectar begins at dusk and continues until late in the morning, although it is most intense at dawn. During late summer and early autumn, when few plants are in flower and nectar sources diminish, the larger and more aggressive females force males out of their territory. If a female is burdened with young too large to ride on her back or in her pouch, she unsentimentally parks them in an abandoned bird's nest or tree hollow until she finishes her foraging.

Honey possums tend to be connoisseurs of those Western Australian wildflowers that grow on small trees or rather stiff, spreading shrubs. Researchers set up small live traps under bushes or within groves, and the marsupials are gently swabbed of pollen before they are released. Samples taken from honey possum fur include pollen grains from various species of clawflowers *(Calothamnus)*, swamp bottle brush *(Beaufortia)*, stick-in-jugs *(Adenanthos)*, Australian honeysuckle *(Lambertia)*, and many banksias *(Banksia)*.

Among the other nonflying Australian pollinators is one species of pygmy possum *(Cercartetus)* in a family (Burramyidae) different from that of the honey possums. Pygmy possums are found from

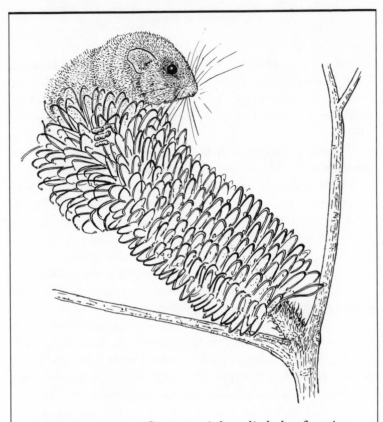

A pygmy possum (Cercartetus) *has climbed a flowering cob of the heath-leaved banksia* (Banksia ericifolia) *to lap nectar from the gutters between the rows of flowers. The hooked "necks" (styles) of the flowers are natural rungs for the grasping paws of these marsupials.*
Illustration by J. Meyers.

the tropical northland to the temperate southland. They are also climbers, and while their habits are not well known, it is clear that nectar and pollen are their primary foods. They are faithful visitors to the southern banksias, and marked pygmies have been recaptured night after night feeding on the same shrubs. Some show a preference for the nectar of eucalyptus trees. Like honey possums, pygmy possums also preen themselves and eat the pollen clinging to their fur. They will consume soft fruits in season, since the teeth of a pygmy are larger than those of a honey possum. Pygmies are active before dawn but continue to feed up to three hours after the sun rises. While all four *Cercartetus* species are known to feed on nectar and pollen, only the eastern pygmy possum *(C. nanus)* has been shown to be a true pollinator of shrubs so far.

The rarest pollinating marsupial now known is the Western Australian dibbler *(Anthecinus apicalis)* from a third family of marsupials, the Dasyuridae. Until nature photographer Michael Morcombe live-trapped two dibblers in 1967, these animals hadn't been collected for eighty-three years and were presumed extinct. No dibbler has been captured since, so almost everything we know about the animal's feeding behavior derives from Morcombe's painstaking notes and photographs.

At the time the dibblers were captured, they were not being studied as potential pollinators. Morcombe was originally trying to catch the more common honey possum by rigging a live trap onto a flowering head of *Banksia attenuata.* When a male and female dibbler appeared instead, he moved them to a large observation cage and recorded that his captives fed eagerly on the nectar of banksia blossoms as provided. In addition, these animals consumed any insects they found on the flowering branches. The dibblers disabled rather large grasshoppers and spiders by buffeting and biting them with sharp front teeth until their prey stopped struggling. They kept odd hours, too, becoming active at sunset for a few hours, sleeping from nine or ten P.M. until five in the morning, and then remaining active until an hour or so before noon.

The pollination of certain plants by flightless mammals has been a compelling theory since the 1930s. At that time German botanist Otto Porsch, having examined the flowers of Australian shrubs, suggested that certain members of the macadamia nut family (Proteaceae) and the myrtle family (Myrtaceae) were especially well adapted to providing nectar to small native climbing rats and indigenous marsupials. Porsch seems to have overextended his theory just a little bit, suggesting that even large kangaroos and wallabies might get into the pollination act as they grazed and poked their muzzles into the tough flowering branches of low-lying shrubs. The evidence for Australian therophily remained more or less circumstantial until the 1970s when fieldwork revealed actual instances of mammal pollination.

Professor Ronald Sussman at the Department of Anthropology at Washington University, Saint Louis, and Missouri Botanical Garden director Peter Raven, have argued convincingly that marsupial pollination was one of the first flower/animal relationships to develop during the evolution of modern plant families. The fossil record clearly shows that 70 million years ago, small marsupials, now extinct, were common throughout the Southern Hemisphere. Botanical studies, meanwhile, have revealed both fossil pollen grains and whole plant remains almost identical to modern banksias (family Proteaceae) and a tree resembling eucalyptus from Eocene deposits estimated to be 60 million years old. The abundance of marsupials parallels the abundance and diversity of flowering plants toward the close of the Cretaceous period (see chapter 8), suggesting a long association between flowers and climbing marsupials even before the rise of the macadamia nut and myrtle families.

Sussman and Raven theorized that marsupials were widespread and important mammal pollinators from the end of the Cretaceous, about 70 to 80 million years ago, until the end of the Paleocene, 55 million years ago. At that time certain groups of marsupials began to fade out in South America and in Africa. Why so many different marsupials became extinct is not understood, but it is possible that nectar-drinking species, at least, may not have been able to compete against the rise and spread of social bees, hummingbirds, and the diversification of eutherian mammals (those mammals that lack a pouch and nourish their embryos through a placenta), such as bats.

Nectar-drinking marsupials have survived only in Australia, with the greatest number of species confined to the woods and shrublands of the temperate south. Flower-visiting bats live in the tropical north. Social bees either never colonized the temperate south or, if they did arrive, became extinct long before European settlers introduced their domesticated honeybees. Some ecologists warn that the commercial honeybee competes unfairly with marsupials, perching birds, and native bees for nectar flows, since beekeepers have transported hundreds of hives into the bushland reserves. Although the honey possum, pygmy possum, and dibbler are the only marsupial species known to actively pollinate the flowers they visit for food, an additional twenty species of marsupials drink nectar at certain times of the year, using this source of chemical energy as a dietary supplement. These more "casual" flower visitors include other tree dwellers, such as the feathertail glider *(Acrobates pygmaeus),* sugar gliders *(Petaurus),* and a shrew-like beastie known as the tuan *(Phascogale tapoatafa).*

Serious field and laboratory studies on the interrelationship between nectar possums and their food flowers are so recent that scientists have split into two different camps. One camp believes that possums are the dominant pollinators of certain plant species. The second group asserts that marsupials are rather undependable pollinators and that those plant species tend to exploit a much wider range of animals to ensure even minimal levels of seedset.

American scientists, including ecologist F. Lynn Carpenter and botanist Delbert Wiens, worked in Australia during the mid 1970s. They concluded that possums may be the most important pollinators of some species in the macadamia nut family, especially certain banksias and their cousins, the dryandras. These shrubs and small trees combine a similar biological clock with specialized physical and chemical features that seem to be adaptations for pollination by wingless mammals. All of these adaptations unite to form a self-consistent "syndrome."

Plants pollinated by possums in Australia produce dense flower clusters on limbs hanging close to the ground. The flowers are often hidden in a maze of twigs and brushy leaves. Individual blossoms are dull rust or straw in color, and the buds open at night. Nectar secretion also occurs at night, with sugary droplets collecting along channels between the floral rows of each cluster. The dull colors and hidden bouquets are believed to discourage foraging by native bees and the perching birds discussed in chapter 3. Possums are attracted to nectar dripping slowly onto the ground at night and to the continuous supply of night blooming buds. The lower the flower clusters hang, the more energy a possum saves, as it does not need to climb very high to find each nectar source.

Since small mammals see tones instead of colors, floral odor is an important attractant, and these flowers release a most peculiar smell, which botanists have tried to describe for decades. Banksias have been said to give off a yeasty smell, but they reminded Porsch of sour milk or caraway liqueur. He found dryandra fragrance reminiscent of cheese and onions. I have given people flowering branches of hairpin banksia *(B. spinulosa)* and recorded their responses. One boy said it was like moldy sawdust. A young woman could not decide whether it smelled more like a clean butchershop or a dog's pelt. Other samplers wrinkled their noses, called it musty, and hastily handed it back to me.

The structure of the flowers has also been cited in support of the "possums-only" interpretation. The pistils of flowers in the macadamia nut family tend to develop swollen, disklike tips at the ends of long, reinforced necks. Pistils are also wiry, flexible, and hooked inwardly, so a climbing animal can use them as ladder rungs. When a possum buries his head in a cluster of banksia flowers to lick up the nectar draining into the channels, the swollen tips of the incurved pistils press against his fur to pick up pollen from flowers visited earlier. A bird's beak, it is argued, is too narrow to come into contact with the curved, angled tip of the pistil, and native bees often have such small bodies that they can land beside the nectar channels on a branch without ever touching the pistil knob.

Carpenter and Wiens acknowledge that there are many *Banksia* species that are pollinated, at least in part, by honeyeater birds. However, they note that banksias more dependent on birds tend to exhibit brightly colored, but rather scentless, flowers displayed on the ends of erect branches. The flower clusters are positioned so that nectar does not drip to the ground, and pistils are straight and stiff. When a bird sips nectar, the knobbed tip repeatedly jabs the feathers of the bird's forehead or crown where pollen grains lodge. Sometimes the knob goes up the nostrils of the bird's bill, which can become packed with creamy yellow grains!

Stephen Hopper, of the Western Australian Wildlife Research Centre, would prefer to deemphasize the importance of proposed syndromes. According to Hopper, there is no such thing as a banksia pollinated primarily by honey possums, or any other wingless mammal for that matter. He asserts that all the members of the macadamia nut family he's examined show a broad mixture of bird and marsupial exploitative devices. There are scentless banksias with hooked pistils and low-growing, smelly shrubs with straight pistils.

Hopper does not feel that other floral features cause a *Banksia* species to lean toward one pollinator or another. Flowers that open rather late at night can fill up with nectar in preparation for the arrival of birds at dawn. He also notes that although the beautiful scarlet banksia *(B. coccinea)* shows every feature associated with a bird-pollinated syndrome, it is still a favorite nectar source of the honey possum. This small tree bears compressed clusters of scentless red and white flowers. The pistil necks are as straight as arrows. These natural bouquets are so brightly colored and long-lived that Western Australia permits their limited collection for the national florist trade.

We are also left with the rather uncomfortable truth that some plants may exhibit all the features of a marsupial syndrome and sometimes grow in regions where marsupials are uncommon or absent yet continue to set good seed due to frequent visits by honeyeaters and feral honeybees. Two scientists recently published results of a two-year field study on populations of heath-leaved banksia *(B. ericifolia)* near Sydney. No pygmy possums were seen or trapped on these shrubs, and the remaining small marsupials never carried *Banksia* pollen. The rust-colored flowers with curved pistils set fruit after visits by birds and bees.

There is some question as to whether marsupials are effective pollinators under the best of circumstances. Since they take a while to climb branches, and they attempt to eat up the pollen that clings to their fur, some researchers feel their contribution to cross-pollination must be negligible. I regard this as nothing less than an act of heresy. Enough trapping and tracking has been done to confirm the movement of marsupials from shrub to shrub and their transport of pollen from one shrub to another.

Vivienne Turner of the Australian National University did her graduate research on the role of pygmy possums in the pollination of coastal banksias. She's had a hard look at the technical literature on paleontology and the comparative merits of bird versus marsupial pollination. Viv suggests that we should not think of the coevolution of birds and marsupials with Australian flowers as an either/or phenomenon. Australian fossils of both parrots and marsupials have been identified dating back to the Miocene epoch (over 30 million years ago). This means that the ancestors of the nectar-drinking parrots (lorikeets) and the honey possums evolved in Australia at much the same time as the modern representatives of the macadamia nut and myrtle families.

Viv suspects that what we've classified as a pure marsupial syndrome really reflects an early association that stressed both marsupials and parrots. Large, compressed flowering branches with long, wiry sexual organs are characteristic of plants also pollinated by lorikeets, who are the only birds known to eat and digest pollen. By the time perching birds like honeyeaters finally arrived in Australia (about seven to ten million years ago), much of the native flora had already evolved floral attractants and rewards that would have been as appealing to true perching birds as they had been to parrots during the previous 20 million years.

It is obvious that the full story of marsupials and their flowers has not yet been told. Moreover, it would be incorrect to suggest that marsupials are the only wingless mammals that pollinate flowers. If placental mammals replaced lines of marsupials in Africa, it's reasonable to expect that placentals have filled marsupial niches vacated by extinction. From a botanical perspective, African members of the macadamia nut family also seem to have become adapted for "Australian style" pollination. Most of the South African proteas we see in cultivation, or in fancy florist shops, are brilliantly colored, nectar-oozing shrubs or trees, with such big, broad flowering heads that they are often mistaken for sunflowers. These species tend to be pollinated by the African sunbirds (Nectariniidae) and perhaps by large scarab beetles.

Other proteas in Africa, however, are low, spreading shrubs with concealed bouquets that are always produced near ground level. *Protea amplexicaulis* and *P. humiflora* are two such species with tight flower clusters that produce a familiar yeasty smell, and the little blossoms are encircled with fleshy bracts that are a dark, reddish maroon in color. During the night these flowers are visited by local rock mice *(Aethomys)*, true rodents that do not appear to be as adept at climbing as the nectar possums of Australia. When these rodents nibble on the sweet and succulent bracts or poke their noses into the cluster of flowers to lap up nectar, their snouts and whiskers are dusted with pollen. The pollen is transferred from shrub to shrub as the protea pistils show the same knobbed tips we've seen in their banksia cousins.

One need not go all the way to the African Cape to witness pollination by rodents. My friend Cecile Lumer was the first field botanist to witness the event in the mist forests of Costa Rica, and she has both specimens and photographs to prove it. The mist forests are found toward the peaks of volcanoes and mountain ranges. Since they receive daily fog baths from low-lying clouds, they can be chilly and damp for a good part of the year. Cecile was studying species of *Blakea,* members of the glory-bush family (Melastomaceae). *Blakea* shrublets are true epiphytes perched on the branches of the mist-forest trees. The more common species produce large, bee-pollinated flowers. A few species, such as *Blakea chlorantha,* offer small, nodding, greenish blossoms that secrete a rich nectar chemically similar to processed white sugar.

These flowers attract climbing rodents, such as the rice rat *(Oryzomus devius),* which visits the inconspicuous blossoms at dusk. Although rice rats have the typical gnawing teeth of their kind, Cecile remarked that they treated the *Blakea* flowers most gently, grasping the flowers with their forepaws and probing the floral bell with their tongues. Pollen is sprinkled on their faces as they lap nectar.

Primitive primates appear in the fossil record shortly after the decline of the marsupials in Africa and South America. Countries such as Peru and Brazil are still home to a number of small monkeys in the marmoset family, known as tamarins *(Saguinus).* When fruits such as figs become unavailable, tamarins depend on the watery nectars of the woody vine *Combretum assimile* and of *Quararibea cordata,* a tall tree. *Combretum* belongs to the Asian wood family (Combretaceae), and its greenish-yellow flowers turn bright orange in color as they grow old and run out of sucrose nectar. Both perching birds and monkeys like the tamarins compete for the limited nectar supply, for it is a major source of calories during part of the dry season. The long stamens and pistil of each flower form a brush, so both the round face and the mustache of a tamarin is lathered with pollen as it takes a drink.

Madagascar is the new "mysterious island" of primate pollination. It still has the world's largest collection of prosimians, especially lemurs, who spend much of their lives in treetops. French and American scientists have observed six lemur species taking nectar from the flowers of at least eleven different kinds of trees and shrubs.

Understanding this relationship involves a sad race against the

Oryzomus devius, *a rice rat of Central America, hangs upside down while it takes nectar from the greenish chalice of* Blakea chlorantha. *Photograph by Richard Schoer.*

clock, as most of Madagascar's forests have fallen to cultivation and subsequent soil erosion. Perhaps it would have been better for this planet had our own ancestors emulated our most distant cousins and remained in the branches as nectar gourmands and tree breeders.

Chapter 7

Beetles, Blossoms, and the Blood of Adonis

And as the dead came back in the sprouting corn, so might they be thought to return in the spring flowers, waked from their long sleep by soft vernal airs. . . . What more natural than to imagine that the violets and hyacinths, the roses and the anemones, sprang from their dust, were empurpled or incarnadined by their blood, and contained some portion of the spirit?

—*Sir James Frazer*

SPRING COMES EARLY to the Middle East and southern Turkey. From late January until the end of May, old pastures and rocky ground are transformed into a carpet of orange and crimson blossoms. Most of these red flowers arise from densely packed colonies of overwintering bulbs and rhizomes, so even the harshest landscape may be vividly colored for a brief time.

This natural pageant was honored by colonies of Jews, who lived in this region after they were expelled from Spain in 1492. These mystics, or Kabalists, celebrated a New Year of the Trees known as Tu B'shvat, which fell sometime between late January and mid-February (using the lunar calendar). On the eve of Tu B'shvat celebrants sampled thirty different kinds of fruits and nuts and drank four goblets of wine. The first goblet was white wine and the fourth was pure red. The second and third goblets were white wine mixed with increasing proportions of red. The Israeli ethnobotanist Nogah Hareuveni interprets this as a representation of the changing seasons in this part of the world. The late summer and autumn landscapes are typified by the bare, gray limbs of almond trees and the white or pastel-tinted flowers of autumn crocus *(Crocus hymenalis)* and white squill *(Urginea maritima)*. These give way to the red tree buds and scarlet wildflowers of winter and spring.

The sight of wild populations of red poppies and tulips is pleasing and reassuring to us. These are the same species that have been used as old breeding stock to coax red hues into many of our own popular cultivars. What would spring in America be, after all, without a bed of red tulips in a city park? The Middle East is also home to a red species of windflower *(Anemone coronaria)*, horned poppies *(Glaucium)*, pheasant's eye *(Adonis)*, and even a giant red buttercup or scarlet crowfoot *(Ranunculus asiaticus)*. In other parts of the world the closest relatives of these Middle Eastern species produce white, yellow, light pink, or royal-blue-and-purple petals. Why should flowers with red pigmentation but almost no discernible fragrance dominate portions of the southeastern Mediterranean?

The many civilizations of the Fertile Crescent were impressed by this red tapestry. They created explanations for this phenomenon and incorporated red flowers into their rituals and folk arts. Their myths often speak of a demigod whose death enriched the earth and whose blood stained the spring flowers. He was known as Osiris, Tammuz, or Attis, the darling *(Naaman)* of the Earth Mother. As early as the seventh century B.C. the Greeks borrowed this myth and gave the doomed deity the name Adonis, who was killed by a boar and whose blood created a red anemone. Sir James Frazer, the great mythographer, felt that such myths were really about human sacrifice. Nature demanded the death of a revered youth or she would not give grain at harvest time.

The Hebraic cultures, on the other hand, associated red wildflowers with joy. Solomon sang, "The flowers appear on the earth, the time of singing has come, and the voice of the turtledove is heard in our land." The king's word for flowers was *nitzanim,* which some authorities feel refers specifically to blossoms with red petals *(nissan* is an Arabic word applied to many red flowers). There is also the questionable identity of Solomon's rose of Sharon (Songs 2:1), which has been interpreted by modern scholars to be the mountain tulip *(Tulipa montana).* "Consider the lilies of the field, how they grow; they neither toil nor spin; yet I tell you, even Solomon in all his glory was not arrayed like one of these (Matthew, 6:28)." It is likely that this gaudy flower is really the windflower known as the crown anemone *(A. coronaria)* with its vivid red bowl, white ring, and velvety black stamens.

Several of the Middle Eastern species display flowers with a crimson rim and a blackened base. They remind people of burning coals. If a Persian boy gave a girl one of the red and black tulips, he would be telling her that his own heart had been charred by the fire of his love.

Explanations for the biological significance of these masses of scarlet have been remarkably recent compared with the several millennia of poetry and theological scrolls. There seems to have been a startling convergence of form and color in at least three different families of plants. Each species exhibits a large, solitary flower at the tip of an erect stalk. The ring of petals or tepals forms a glossy, red bowl that lacks a distinct floral perfume. The flowers of all these species lack nectar glands.

Red, odorless flowers are most commonly pollinated by birds. However, the nectar-drinking sunbird *(Cinnyris oseae)*, which is distributed through the Middle East, ignores the spring display of red bowls. When I was a student, I would have explained this anomaly by saying that the red flowers weren't really red. A combination of magenta, mauve, and pink pigments suspended in the water bags (vacuoles) of the epidermal cells of the petals produced an "illusion of redness" to the human eye. Unfortunately for this neat explanation, the wavelength of petal colors can be measured with a device called a spectrophotometer, and the Middle Eastern species all fall within 600 to 800 nanometers, well within the true orange-red range.

A team of seven scientists from Israel, England, and the United States have come to the conclusion that beetles are the scarlet-flower fanciers in the Middle East. Specifically, there are about half a dozen species in the genus *Amphicoma* that spend much of their adult lives transporting the pollen of red flowers. *Amphicoma* beetles belong to the Scarabaeidae, a huge family of insects that have been treated with a mixture of awe and revulsion throughout the ages. The sacred scarab *(Scarabaeus sacer)* of Egypt is one of many tumblebugs that roll balls of dung on which they lay their eggs. Skin beetles (Troginae) are the bane of tanners and furriers, since they devour dry animal carcasses. No one is very fond of the Japanese beetle *(Popillia japonica)*, which turns our gardens into a salad bar, or the flower beetles (Cetoniinae), which attack ripening fruit. Other members of the family consume carrion, fungi, and decaying vegetable matter.

Amphicoma species do not waste their days gathering dung or gnawing rotting logs. They appear to dine exclusively on the pollen produced within red flowers. The beetle's body is covered with dense bristles, which trap spilled pollen. When the pollen-dusted insects scramble over the sticky, receptive pistils, some transfer is made, and a new generation of seeds starts to develop. (See color plate 5.)

The beetles also use the open floral cups as mating sites, which is why a male *Amphicoma* is a superior pollinator compared with the female. Female beetles may remain in a floral cup for fifteen minutes or more grazing placidly on pollen. A male cruises his habitat for food and females, so he changes flowers almost every three minutes until he locates a suitable bride.

This phenomenon is far more common than most people assume. Beetles pollinate many tropical plants, including the palms, philodendrons, Panama hat plants, and some of the magnolioid trees and water lilies I will discuss in chapters 8 and 9. However, these flowers are often praised more for their lush perfumes than for their dull cream, greenish-white, or maroon colors, and beetles are better known for their remarkably developed sense of smell than for their vision. Do *Amphicoma* species really seek out red flowers or are they attracted to a subtle aroma that eludes humans?

To test the beetles' color appreciation, Dr. Amots Dafni, of the University of Haifa at Mount Carmel, devised a simple but cunning experiment. Dr. Dafni mounted a series of colored cups (red, blue, yellow, green, brown, and white) on foot-high sticks (the average height of a red flower stalk) and set them out in the spring in a field filled with real red blossoms. Dr. Dafni and his assistants recorded *Amphicoma* species visiting the unscented cups 136 times over twelve hours of observation. Interestingly enough, 87 percent of the beetles were found only in the red models. The remaining 13 percent were distributed almost evenly among the other cups.

How this unusual relationship between beetle and blossom evolved is still not understood, since so little is known of how any beetle perceives the world through its compound eyes. *Amphicoma* species do not seem to care much about the smell of a flower. They just fly to a red object; they have little regard for its size or depth. Dr. Dafni found that beetles would even dive into a child's red pail, the sort used for fun at the beach, if it was left upright in a field.

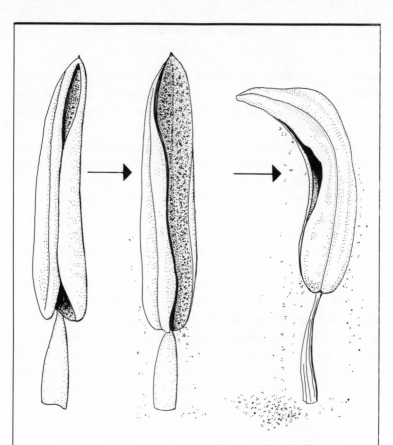

The ripe anther of a wild tulip offers pollen to beetles by "unzipping" two slits (each lying on opposite sides of the same anther) and exposing the contents of the pollen sacs. The drying anther bends and twists, wringing out pollen grains that collect at the base of the floral cup like piles of blackish soot. Illustration by J. Meyers.

This beetle pollen banquet lasts over four months and is served as a series of overlapping courses, each becoming more generous as the season progresses. The crown anemone flowers during the last week of January, but each flower contains less than five hundred pollen grains. The red poppy *(Papaver carmeli)* does not pop open until early April, but each blossom dishes up more than five thousand grains. Consequently the majority of *Amphicoma* species aren't found until March. By then there are enough tulips and scarlet crowfoots to sustain the beetles until the generous dessert of poppy pollen is served a month later.

Most of the red flowers are crammed with over four dozen stamens. Each is tipped by an anther head, which spills pollen all over itself and turns each stamen into a pollen lollipop. Tulips have only six stamens per flower, but they manage to pack more pollen grains per stamen than any other red flower. Even our cultivated tulips release copious quantities of pollen the same way their wild relatives throughout the southern Mediterranean do. The arrowhead anthers split longitudinally and twist as they dry out, wringing piles of pollen onto the marked spots at the base of each petal.

It's only after the great stores of pollen from red flowers have been depleted that the beetles travel to other species. By the last couple of weeks in May, *Amphicoma hyrax* and *Amphicoma genei* can be found on the blue heads of the knapweed *(Centaurea cyanoides)* and other white or yellow flowers in the daisy family. I'd like to compare these daisies to afterdinner mints, but you would probably groan at that simile, and then your open mouth would be like a red flag to a beetle.

Chapter 8

An Ancient Moth for an Ancient Flower

And I have by me, for my comfort two strange white flowers—shrivelled now, and brown and flat and brittle—to witness that even when mind and strength had gone, gratitude and a mutual tenderness still lived on in the heart of man.

—H. G. Wells

THE ORIGIN OF flowers may best be regarded as a classic and elaborate "whodunnit?" Each of the previous three chapters indicates that variation in floral advertisement and form reflects a continuous accommodation by plant populations to a changing cast of foraging animals. When scientists debate the equipment and architecture of the first flowers, they are really discussing how animals influenced the design of a prototype that ultimately improved the efficiency of seed production. That basic design was so successful that it still dominates most of the landscape on this planet, although it has undergone many radical modifications.

How can we search for these first flowers when scientists estimate the plants that produced them have been extinct for over 120 million years? We must begin by examining the work of paleontologists, who have made some exciting but isolated discoveries of fossilized insects and floral parts over the past two decades. Combine these findings with the fact that some living families of both flowering plants and insects with remarkably old lineages look very much like the fossils. Perhaps by studying these living descendants in their natural habitats we can build up a model for prehistoric pollination. Botanical sleuths' hypotheses are balanced: the hard evidence of picks and shovels combined with the contents of a butterfly net held under a hand lens.

The winter's bark or pepperleaf family (Winteraceae) offers some of the oldest recognizable fossils. Recently, Drs. James and Audrey Walker, at the University of Massachusetts, and Gilbert Brenner, from SUNY at New Paltz, examined core samples taken from the northern Negev of Israel. They discovered the pollen of Winteraceae in deposits that could be dated back to the early Cretaceous. The significance of fossilized pollen clusters that are 105 to 110 million years old cannot be underestimated. Obviously the Winteraceae were among the first trees and shrubs derived from plants with primitive, flowerlike organs.

How did this consortium of scientists know they had found

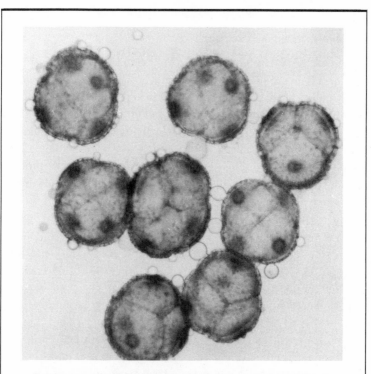

Pollen tetrads of Zygogynum bicolor *linked together by clear droplets of oil (pollenkitt). Each tetrad consists of four pollen grains fused together by a common inner wall. Each grain bears a circular "nipple pore." The coat of oil allows the greasy tetrads to be extruded from their stamen like toothpaste out of a tube. Photograph by P. Bernhardt.*

pollen of the Winteraceae? The two-layered wall of these pollen grains has a unique configuration that distinguishes it from all other pollen forms either living or extinct. The fossil grains formed interconnecting walls, so that each pollen "unit" was actually a cluster of four grains (tetrads). Each individual grain had a bumpy, pitted surface with a single, reinforced pore that resembled a little nipple. Only a few living genera in the Winteraceae still produce pollen walls combining all three characteristics. The only minor difference between living and extinct pollen clusters is that the pits on the surface of modern grains are wide enough and deep enough to form microscopic craters, known as lacunae.

Over the past 100 million years the distribution of the Winteraceae has contracted severely. They are extinct in Israel and throughout most of the Northern Hemisphere. You need to take a trip to the tropics to see winter's bark in any great numbers.

New Caledonia, still a colony of France, is an island that lies 750 miles northeast of Australia. It is filled with both primitive woody plants and certain insects. Under the dense, moist canopy of the rain forest the winteraceous tree *Zygogynum* and the moth *Sabatinca* are locked in an unusual interrelationship.

Sabatinca moths are in the family Micropterygidae, which consists of about one hundred species distributed throughout various parts of the world, including North America. However, entomologist George Gibbs, of the University of Wellington in New Zealand, states that twenty-six *Sabatinca* species make New Caledonia their home, and the island has the largest *Sabatinca* diversity. It now looks like these moths have a pedigree as old as, or older than, the winter's bark family.

Amber found in Lebanon estimated to be 100 to 130 million years old contains the body scales, limbs, and whole bodies of moths identified as micropterygids. In fact, the anatomical remains are so similar to living *Sabatinca* moths that the extinct insects have been named *Parasabatinca*. The remains of other micropterygid moths have been identified in Baltic amber but are no more than 40 million years old. This suggests that micropterygid moths, like the winter's bark trees, have suffered a declining range over millions of years.

Zygogynum trees produce many flower buds, but only a few open each day. Consequently one small tree may remain in flower for months, despite the fact that each blossom lives no more than two days. This is a common flowering strategy of many tropical trees, as we saw in chapter 1. Each flower usually has two rings of stiff petals. The petals of some species have a pinkish tinge while others are yellowish-orange or a deep burgundy.

The pollen-making stamens and seed-making carpels live together in the same *Zygogynum* flower, but they mature at different times. When a flower bud opens its petals promptly at 7:00 A.M. on the first day of its life, the carpels are receptive and ready to be pollinated, but the stamens stay shut. When the same petals expand on the second and last morning, the pollen-receiving tip (stigma) of each carpel is dried up and crusty, but the stamens begin extruding their pollen. By isolating the sexual organs from each other in this way, *Zygogynum* trees avoid accidental self-pollination. When carpels mature and collapse before the stamens, it is called protogyny (*proto* = first; *gyny* = female). Protogyny was probably developed to encourage cross-pollination in the first flowers. It is found extensively throughout the winter's bark family and in many more relictual plants with primitive floral organs. This includes the protogynous blossoms of the magnolias *(Magnolia)*, custard apples *(Annona)*, star anise *(Illicium)*, and many of the water lilies, including the huge plant of the next chapter.

The stamens of many primitive flowers extrude their pollen much like cone-bearing plants (gymnosperms) such as pines *(Pinus)* or cycads *(Cycas)*. Such trees produce male cones that shed great quantities of pollen. This pollen extrusion also suggests that primitive blossoms tend to be pollinated by insects lacking limbs designed to grasp and manipulate the sexual organs of flowers. Such insects can only lap pollen from collecting troughs at the flower's base or from floral structures covered with it. The stamens of a *Zygogynum* flower extrude pollen most generously. When ripe, a short pair of slits opens along each stamen tip, and pollen is literally squeezed out like toothpaste from a tube.

The pollen tetrads dangle in sticky strands because pollen clusters are embedded in an oily matrix called pollenkitt (pollen mortar), a thick, greasy material that keeps the strands coherent. We now see that the bizarre craters on the wall of each pollen grain expand the surface area of each grain so that it can accumulate dense deposits of pollenkitt.

Zygogynum flowers offer no nectar. Nectar glands appear to be a comparatively advanced feature of modern flowers. This absence does not appear to trouble *Sabatinca* moths. They lack the equally advanced long, strawlike tongues associated with butterflies and sphinx moths (see chapter 16). The *Sabatinca* moth has mouthparts capable of scraping and grinding food, however. The caterpillar passes its youth gobbling mosses, liverworts, and the refuse of the forest floor. As a winged adult its tastes become slightly more elevated as it chews up the spores of fern trees and the hard-walled pollen of flowering plants. These moths are quite pretty for their small size. They're luxuriously furred and have "bunny rabbit" faces. Their fringed wings exhibit contrasting zebra patterns that are often accented by brighter colors.

Swarms of *Sabatinca* moths literally strip a *Zygogynum* flower of every greasy thread. A few pollen clusters escape the moths' mouths and adhere to their legs and wings. The oily gunk dries to a white shellac, which is wiped off onto wet stigmas when the moth explores a first-day flower in search of more pollen threads.

Of course, a first-day flower offers no pollen food, but *Sabatinca* moths may not visit a flower just to fill their stomachs. Once enough moths build up on a flower, some species initiate a dance, crawling in a counterclockwise, looping circle. It ends in copulation, and it appears that a first-day (stigma-receptive) flower is as suitable for an orgy as a flower dripping with pollen.

Other *Sabatinca* species don't even bother to dance but copulate aggressively when the sexes meet. During their brief courtship and coition the moths may pass over a receptive stigma and deposit a pollen cluster. Think of the *Zygogynum* flower as a "living

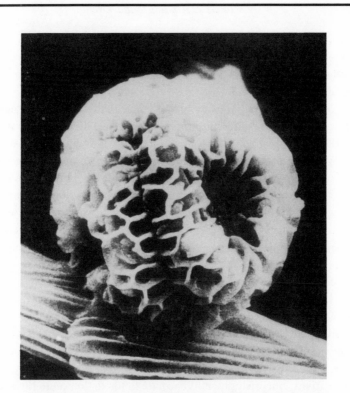

A tetrad of Zygogynum *pollen grains has adhered to the leg scales of a* Sabatinca *moth. Two interconnected grains face the viewer, displaying the entrances to the dark pores and the honeycombed ridges on the outer pollen wall (exine). Deposits of white, gunky pollenkitt can still be seen in the exine "craters." Photograph by P. Bernhardt.*

motel." Moths are guaranteed a place to meet and perhaps something to eat. Payment for the tree is successful seedset and/or pollen transport.

Sabatinca moths are tiny, usually less than ¼ inch in length. How do they find *Zygogynum* trees in the dense, diverse understory of the rain forest? For that matter, how do they find each other? They are surprisingly efficient on both counts. Swarms of more than thirty moths have been observed on a single *Zygogynum* flower.

The answer lies in the fragrance offered by the flowers on both days of their brief life. So strong is the scent that moths can even detect the fragrances of green buds just as the petals begin to crack apart. The flowers of some *Zygogynum* species produce a musty smell, but others produce a pungent perfume that has been compared to burned orange.

Leonard Thien of Tulane University collected the fragrances of *Z. bicolor* and *Z. baillonii* using a special absorptive powder developed in Sweden. Samples were sent to fragrance experts at the Universities of Uppsala and Göteborg, where thirty-six compounds were identified. They include terpenoids, benzoids, and fatty-acid derivatives. Surprisingly, ethyl acetate tends to dominate the spectrum of aromas.

Most of us know ethyl acetate as the loathsome chemical that, in its concentrated form, can damage the mucous membranes of human noses when inhaled. It is also quite fatal to most insects. Wise collectors prefer ethyl acetate to cyanide fumes in their killing jars because it kills large insects, such as dragonflies, quickly without leaching them of their natural colors. What is this deadly poison doing in an insect-attracting scent? Leonard Thien told me recently that shortly after he published his results, he received "hate mail" from an entomologist who accused him of faking his findings.

As a weak, diffuse gas, ethyl acetate smells rather tempting, like poached pears or dried apples. You have probably eaten it, for it is often added to cheap candy and canned food (in safe proportions, of course) to give them fruitier odors. Weak concentrations, further diluted by thirty-five other compounds, may act as a powerful draw to a small insect that perceives the world primarily through its superior sense of smell.

It's possible that female *Sabatinca* moths lack the ability to release sexual odors (pheromones) needed to attract virile males. Pheromones are a common adaptation of more advanced moths. However, by exploiting the fragrance of *Zygogynum* flowers as an "assembling scent," these tiny insects can locate each other and food sources as well in the tropical forest.

All the components of *Zygogynum* floral fragrance are also found in the tissues of their green leaves. This is not too surprising, as all flowers are really rings of folded leaves, modified for seedset. High concentrations of ethyl acetate in leaves act as a natural repellent driving away insects that eat leaves. This is recognized as an important stage in the evolution of coadaptation between the first flowers and their prospective pollinators. Systems of chemical defense evolved into a syndrome of mutual benefit to plant and insect.

When a pollen cluster lands on a wet stigma, one pollen tube emerges from each of the four nipple pores. If each carpel of a flower is fertilized with the sperm from one pollen tube, the carpel ovaries fuse together to form a woody fruit about the same size and density as a golf ball. Fruit development is rather slow (another common feature of primitive trees), taking almost a year. We're not sure how the seeds are released, but it looks as if the fruit simply drops off the twig, and the outer fruit wall rots open with age.

Have micropterygid moths been feeding and fornicating on winter's bark flowers for the past 110 million years? The fossil evidence certainly entertains this possibility. In the early Cretaceous, both Israel and Lebanon would have been located on the equator and would have been a part of the supercontinent Gondwanaland (which consists of the rafting tectonic plates of Africa, South America, India, Madagascar, Australia, and New Zealand). This is the ancient region where the first flowering plants are believed to have evolved.

Leonard and I think the evidence is so convincing that we are willing to say the ancestors of modern *Sabatinca* moths were at least one of several pollinators of the first flowers. While *Zygogynum* is pollinated by moths, it must be noted that other members of the winter's bark family are serviced by other insects. The *Exospermum* trees of New Caledonia produce pink, fruity-scented, urn-shaped flowers that are pollinated only by beetles. The pepper-leaf bushes *(Tasmannia)* of Australia and New Zealand depend primarily on small flies. It seems most likely that these living relicts all spring from a common ancestor that catered to several different kinds of insects that ate extruded pollen. It's just that plate tectonics and the fossil evidence may currently provide better "press" for the ancestors of *Sabatinca* than for modern blossom beetles or flower flies.

Of course, fossils are not action photographs developed in stone. No one will ever find a stone bee in the act of picking pollen off of a petrified geranium. Reconstructing the history of flower pollination is like rebuilding a castle from a collection of old keys. *Sabatinca* and *Zygogynum* have become one of the most important lock-and-key sets in a 120-million-year-old mystery. Further exploration of the rain forests may reveal even more archaic flowers that have been sculpted by a moth's mandibles.

Part Three

Portraits in Chlorophyll

Chapter 9

Royal Amazons

A fair imperial flower;
She seemed designed
for Flora's hand,
The sceptre of her
power

—William Cowper

VICTORIA IS THE sort of name that is associated with things on a grand scale. If you don't care for the state of Victoria in Australia, you can visit Victoria Falls in Africa. The Victoria Cross is awarded only for acts of exceptional valor, and the Victorian era was the high point of the British Empire. *Victoria* is also the "Christian name" of a genus of water lilies that are the leviathans of their family. These tropical plants produce floating lily pads up to 12 feet in circumference and flowers as massive as heads of cabbage.

Western civilization was properly introduced to giant water lilies in 1803 when the Czech botanist Thaddäus Haenke, discovered them in Bolivia. It is said he was so overcome by the sight of the plants that he fell to his knees. For the next thirty-five years British, French, and German explorers found the giant plants in the rivers of Brazil, Argentina, and the Guyanas. They wrote enthusiastically of the almost supernatural beauty of the foliage and blossoms and then attempted to procure specimens for the botanical collections of Europe.

In 1833 the German botanist Eduard Poeppig received enough preserved material to describe a new species. He recognized that the specimens belonged to the water lily/lotus family, but he thought they were just giant versions of prickly water lilies, so he named them *Euryale amazonica*. In 1837 John Lindley examined specimens from the dried-plant collection of Kew Gardens. Because Princess Victoria had ascended to the throne that year, loyal Lindley dubbed the plant *Victoria regia*. The generic name, *Victoria*, has stuck thanks to Lindley's meticulous descriptions of the unique floral features. Today the taxonomic confusion is over, and we recognize two species. *Victoria amazonica*, from hot, equatorial South America; and *V. cruziana*, from cooler climates as far south as Argentina.

For collectors, the problem has always been how to introduce these royal amazons to cultivation. For more than forty years, attempts to grow plants from seed or transplanted rootstock failed. Even the clever French, so adept at taming the plants of China and Australia, were unsuccessful. It was not until 1849 that live seeds survived the trip to England. Two British physicians living in Demerara, Guyana, realized that victoria seeds must remain wet to stay viable. Their solution was simplicity itself: send the pea-sized seeds in bottles of clean water!

The seeds sprouted easily at Kew Gardens, but then the director, Sir W. J. Hooker, realized he had a surfeit of riches. With Kew's limited facilities, how could all the plants reach maturity?

Plantlets were farmed out to some of the great British estates, and a few lucky little victorias went to Chatsworth, the estate of the sixth duke of Devonshire, who was a great patron of tropical horticulture. The duke employed Joseph Paxton (1801–1865), an inspired gardener known for his revolutionary techniques for coaxing temperamental exotics, who realized that tropical plants flourished only when their environmental conditions were duplicated in captivity.

Paxton had a tank built for his charges that was 12 feet square but only 3 feet 4 inches deep. One plantlet was given a submerged bed of rich soil. A little waterwheel was installed to duplicate the sluggish current of a South American river. The baby victoria responded to its new home with such rapid, intense growth that the pool had to be enlarged. Within a little more than three months after its installation, the royal amazon was in flower. On November 13, 1849, a massive leaf and a beautiful blossom were presented to Queen Victoria at Windsor Castle. After viewing her namesake, the monarch announced, "We are immensely pleased."

Paxton's fascination with his victorias did not end with his horticultural triumph. He studied the development and structure of the plant intently and uncovered some startling secrets. The water lily could grow 647 square inches in a single day! When the pad was mature, its edges curled inward giving the leaf the appearance of a shallow canoe. When taken from the water, the leaves were soft and flabby, but when allowed to float, the leaf retained its symmetry. (See color plates 6 and 7.)

Paxton soon realized that the leaves were held together by tough, articulated veins. These veins did more than circulate food and water, though. They were also filled with air passages, arranged along the underside of each leaf, producing a durable source of buoyancy and reinforcement.

On November 22, 1849, Paxton dressed his seven-year-old daughter, Annie, as the queen of the fairies and placed her on a leaf in the victoria pool at Chatsworth. The leaf supported her weight with ease. It has since been shown that some large victoria leaves can support up to 300 pounds of dead weight without capsizing. Paxton learned an important and profitable lesson from this piece of natural engineering. He incorporated part of the cantilever system of the lily pad into his design for the Great Exhibition Building of 1851. Critics warned that his plan for reinforcement would never support so many panes of glass. A water lily proved them all wrong, and the building became known as the Crystal Palace, a wonder of Victorian architecture. Queen Victoria eventually knighted Paxton for his contributions to architecture and industry.

Once the requirements for victoria culture were published, most of the great botanical gardens of Europe responded with enthusiasm, sophistication, and a touch of whimsy. Special houses were constructed in the gardens of Ghent, Brussels, Berlin, and so forth, specifically to exhibit victorias year-round. Attempts were made to transform the interiors of these water lily houses into romanticized portions of the Amazon basin. Lavish landscapes encouraged interest in other exotic aquatics grown with the victorias.

Victorias did not reach the American public until almost the end of the last century. James Gurney, the head gardener at the Missouri Botanical Garden in Saint Louis, had spent part of his youth working for Paxton on the original project to grow the first royal amazon. Gurney successfully introduced victorias to Saint Louis, and the first plants flowered in 1894. They have been an annual attraction ever since. There is a series of delightful old photographs featuring children, elegant ladies, policemen, and horses poised on victoria leaves.

The first collectors were ensnared by the flamboyant flowers. Prussian botanist Sir Robert Schomburgk wrote, "I rowed from one to the other and I always observed something new to admire." The French botanist Aimé Bonpland nearly fell off his raft in a feverish attempt to gather flowers. The Tupi Indians of South America understood this obsession. They knew that victoria flowers contained the soul of a chieftain's daughter who had drowned herself so that she might end her passion for the aloof god of the moon.

Scientific studies of victoria flowers have generally kept pace with work on the vegetative organs. They provide invaluable information on the origin and evolution of pollination systems in the tropics. Ghillean Prance, of the New York Botanical Garden, and entomologist Jorge Arias, of the Instituto Nacional de Pesquisas, studied wild plants of *V. amazonica* in Manaus, Brazil, and published a definitive paper in 1975.

The water lilies bloom throughout the year, but each huge flower lives only two days. The white bud of youth blushes pink with age until it reaches the red-purple of senility after twenty-four hours.

On its first day of life the flower opens as dusk falls. The smell is strong, reminiscent of butterscotch mixed with ripe pineapple. The interior of the flower is usually warmer than the surrounding air, due to the breakdown of starch granules in floral cells. During this first evening of life the flower's pistils are receptive to pollen, but the anthers (pollen-making organs) lie flat and shut. As dawn approaches, both the floral temperature and odor wane, and the petals close.

The flowers reopen on the second day between 10:00 A.M. and noon. The pistils are withered, but now the anthers split open and pollen is scattered over the floral face. By the end of the day all the floral organs collapse and the dead bloom is no more attractive than a rotten cabbage.

This display of floral movement, color, fragrance, and tempera-

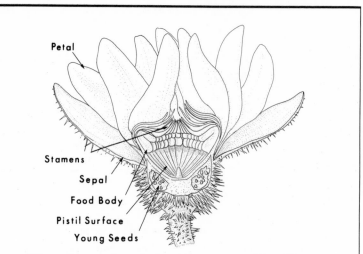

Petal

Stamens

Sepal

Food Body

Pistil Surface

Young Seeds

When the petals of the giant water lily (Victoria amazonica) first begin to expand and separate, the sexual organs have already formed a chamber and restaurant for their beetles. The beetles gnaw on the spongy food bodies or copulate on the receptive surface of the pistil.

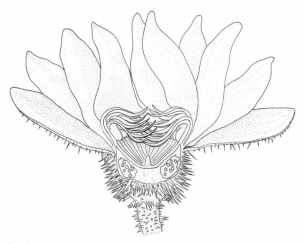

As the flower ages, the pistil surface loses its receptivity, but the stamens expand to shroud the food bodies and to shed their pollen on the beetles. Illustrations by J. Meyers.

ture change is for the benefit of the large scarab beetles *Cycloceph-ala*, which crowd into the flowers at night. The closing petals do not distress them much, and the flower is warmer than the surrounding atmosphere. Inside the floral chamber the insects dine on starchy food bodies, which lie above the pistil surface, and exploit the protection of the closing flower to search for mates and to copulate. Recent experiments have shown that fat scarabs run the risk of being gobbled up by sharp-eyed birds or monkeys if they mate in plain view.

When the flowers open on the second day, the beetles are dusted with pollen before they fly away. They carry the pollen to the next first-day flower they visit and may pollinate the pistils as they search for more food bodies. It's extremely rare for a flower to pollinate itself.

If a flower has been fertilized, it sinks into the river. When the massive pod matures and decomposes, the seeds escape and float back up to the surface, where they are dispersed by the current. Since Brazilian rivers tend to flood annually, a seed may be carried a considerable distance from its parents.

The life span of a victoria revolves around the life of the tributary in which it lives. Some royal amazons survive for only one glorious season of growth. When their branch of the river dries up, they die, which means that some huge plants last just a little longer than a petunia or a marigold. Those victorias inhabiting permanent stretches of water may, of course, survive indefinitely, growing bigger every year as their root systems expand in the mud.

Most institutions lack the space and facilities to breed victorias year-round. Longwood Gardens in Pennsylvania, however, has become the center for seed production in North America. Their indoor facilities are sufficiently large, and their internal heating systems are so superior that they can grow *V. amazonica*, the most finicky of the two species. *V. amazonica* refuses to flower freely unless its "bath water" is at least 80 degrees Fahrenheit, and they really prefer temperatures of 85 degrees and higher.

Longwood has pioneered advances in victoria husbandry by selective breeding. In 1961 Patrick A. Nutt produced the first hybrid between *V. amazonica* and *V. cruziana*. The offspring combined the temperature tolerance of one parent with the robust growth and color of the second. Summer displays of victorias, confined to outdoor pools in city parks and arboreta, tend to be dominated by these remarkable hybrids from Longwood.

During a recent visit to the Missouri Botanical Garden, I met Mary Smith, a self-described, "woman-of-all-work" to six victorias. Every year the Garden places two cruzianas and four hybrid victorias on display in a circular reflecting pool in front of the famous Climatron. There is surprisingly little glamor in the life of lady-in-waiting to clusters of queenly water lilies. To achieve an exquisite display requires precise planning, due to the peculiar life cycle of the victorias.

Bottles of seeds arrive from Longwood Gardens each winter, and Mary pops them into the fridge at 40 to 50 degrees Fahrenheit, since low temperatures delay germination until the garden is prepared. To sprout victorias, each seed is given its own 4-inch pot which is then submerged in a tank of water maintained at a constant 85 degrees. The seeds send up their first leaves after a week. Within a month the sturdy plantlets are moved to 7-inch azalea pots until it is warm enough to transfer them outside. The tender plantlets can suffer from attacks by shrimplike animals called water pennies that burrow into their soft tissues, so the tanks are patrolled by schools of hardy blue paradise fish, which lunch on any crustacean or insect intruders.

Now comes the messy part. I've watched Mary don her hipwaders, then empty and scrub the reflecting pool each spring before transplanting begins. One victoria is placed in each of six planters arranged around the pool. Transplanting requires protection for the horticulturist. Victoria leaves are armed with long, sharp, jagged prickles. Of course, the plants can't go outside until the water temperature remains at a constant 70 degrees. This usually means that the victorias do not leave the greenhouse tanks until the end of the second week in May.

Humid Missouri summers can induce flowering by early July. Of course, the plants become so huge that they can't be transferred back to the greenhouse at the end of the growing season. It's far easier to let the frost kill them each autumn and buy new seeds every winter. By early October I claim victoria leaves for the students taking my course in General Botany. The cantilevered system of veins is worth examination during lab sessions, although there are often shrieks of agony from curious botanists who have punctured their thumbs on the prickles.

Victorias suffer from two pernicious pests at the Missouri Botanical Garden. Aphids attack the upper surface of the lily pads, but they can be treated with a mild insecticide. People, though, can not be treated in the same way. Mary complains that victorias suffer from the hot-penny trick. Visitors like to test their aim by pitching pennies onto the pads. On hot days the coins heat up and burn through the plant tissue, leaving ugly black-and-yellow-rimmed holes. Mary has had signs erected that state, "Your wishes will come true if they land in the water and not on the lily pads."

The victorias remain a popular attraction at the gardens. Visitors congregate around the pool, pointing at the pontoon pads, which rock from side to side under the slightest breeze. People are surprised that one plant may bear flowers with several different colors. Mary tells me that many are amazed that such giants are actually grown from seeds. Even the mallards that nest in the Garden can be seen late in the afternoons taking their new ducklings for a spin around the lily pads.

Evolution may have made these plants queens of the tropics, but mankind has kept *Victoria* on the throne of popularity for 125 years. This has come about only because a few people have taken the time to determine how the plant has adapted to its original environment and how such conditions can be exploited. What a pity that Charles and Di do not thrive under similar scrutiny.

Chapter 10

The Wily Violets

The best I can find to say of these coarse
rampageous Violets is that they will thrive
anywhere and make unobtrusive masses in
any cool, good soil.

—*Reginald Farrer*

THE PHRASE *SHRINKING VIOLET* is really an over-worked oxymoron. Violets have not inherited their corner of the earth due to any intrinsic meekness. Research, begun in this century, has shown continually that these small yet aggressive plants deploy a number of reproductive tactics to ensure survival.

Although violets are well established in the high-altitude regions of both Old and New World tropics, the largest concentration of the five hundred species of the genus *Viola,* in the family Violaceae, are in the Northern Hemisphere, and nearly 20 percent of all *Viola* species live in the United States. Most American violets are native species, except for a few naturalized Europeans, fugitives from gardens or stowaways in sacks of grass seed.

While they show some preference for consistently moist soils and semishady conditions, violets are common to nearly all major habitats. At least one species can be found in virtually every swamp, prairie, deep forest, or meadowy dell, flowering away from late February through mid July (depending on latitude and climate). Some, like the common blue violet *(Viola papilionacea),* thrive where man has disrupted the earth and have come to be regarded as just another suburban weed.

Attempting to permanently excise blue violets from a wet, acid lawn is a lot like plucking the proverbial gray hair. Tear out a clump, and ten more grow in its place—weeders forget that quite frequently important plant organs are not visible to the human eye.

The vigorous entrenchment of the common blue is characteristic of other violets and many herbaceous perennials as well. Violets lack an aerial stem and seem to exist as independent clusters of leaves punctuated by flowers on stalks. The true stem, however, is a short rhizome, a creeping horizontal structure that grows underground. Like most rhizomes, those of the "stemless" violets show a unique ability to regenerate. When violets are pulled out of the lawn, there are often a few pieces of the branching rhizome left in the soil to start new plants. Violet patches in abandoned gardens may consist wholly of clusters descended from a single original rhizome. These natural clones, from vegetative propagation of stem pieces are not unique to greenhouses. They are found frequently in nature.

Vegetative propagation is only one device in a violet's arsenal of survival strategies. Most violets bear not one but two radically different types of flowers. The showy or chasmogamous flower is the sweetly scented, handsomely colored blossom that has made the genus *Viola* so popular. These flowers sport olfactory and visual cues that indicate they're insect pollinated.

British-born botanist Dr. Andrew J. Beattie has explored the sex lives of North American and English violets, and much of what we know of their pollination ecology is due to his research. While older naturalists believed that only a few species of solitary bees fertilized the pistils of showy flowers, Dr. Beattie has determined that bumblebees, wasps, sawflies, hoverflies, dance flies, and March flies, as well as some hawkmoths and butterflies, play significant roles in the transport of violet pollen.

Dr. Beattie has found that violets exploit such a large variety of prospective pollinators because violet flowers are long-livers and shape-shifters. Compared with the flowers of other spring plants, violet flowers last a long time. The showy blooms of the British wood violet *(V. riviniana)* and the hairy violet *(V. hirta)* last from one to two weeks and are remarkably resistant to periods of freezing. As Byron noted,

> With us she is but a winter flower
> The snow on the hills cannot blast her bower.

In addition, during its lifetime the flower's stalk changes its position several times, and the flower's face assumes nodding, then horizontal, and finally erect positions. The petals also move gradually, so that at different times in its life it will be able to service pollen-eating flies and nectar-sucking bees and butterflies.

Yes, there is a sweet, liquid reward for insects with tongues long enough to reach the prize. Most violets wear a succulent nectar

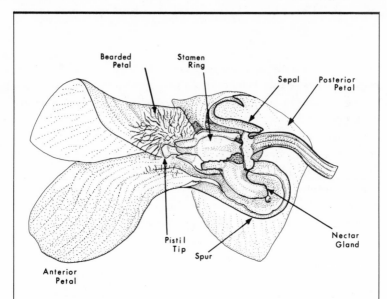

The interior of a woolly blue violet flower (Viola sororia), *which invades so many wet, suburban lawns. The anterior petal forms a hollow spur at its base to catch the nectar droplets secreted by large nectar glands attached to the lower stamens. To reach the store of nectar, an insect must push its head under the receptive pistil tip (stigma) and the ring of pollen-laden stamens. Illustration by J. Meyers.*

gland on two of their stamens. The nectar drips into a hollow pouch formed at the base of the specialized "spur petal." To reach the collected droplets of nectar, an insect must insert its tongue under the pistil and the ring of stamens in order to probe the contents of the pouch. The way in which pollen is given to a thirsty insect depends, in part, on the size and shape of the floral organs but also on the foraging "tricks" employed by different pollinators.

For example, flies and butterflies prefer to feed on nectar in a prone or horizontal position, so they are most successful on a *Viola* flower that has a spur petal with a wide, anterior rim that can be used as a landing platform. Pollen is deposited dorsally on the insect's "forehead and back" while it sucks nectar. Small bees, on the other hand, appear to be comfortable only when they can visit a violet flower while standing on their heads. They land on the rim of the spur petal, but then they rotate their bodies 180 degrees, clasping the two posterior petals with their hind legs. Pollen is deposited ventrally on the bee's "chin and chest."

How do so many different insects "know" where to search for nectar in a violet flower? Why aren't they reluctant to place their heads down into a dark floral tube? Neither innate knowledge nor bravery has much to do with a bee's success in its search for nectar. The violet flower offers a code of color and perfume that directs the pollinator to its reward.

The rim of the spur petal and the entrance to the floral tube are usually well marked with deeply contrasting pigments. Sometimes these floral guideposts consist of converging spots against a much lighter backing or a series of dark, spidery veins or streaks. These patterns are the nectar guides, which direct the foraging insect toward the position of the concealed spur. In some violet species the nectar guides are lustrous and greasy, so the flowers appear to be "freaked by jet," as Milton wrote.

Nectar guides have intrigued and delighted people of diverse cultures for thousands of years. Some have interpreted the alternating patterns of orange, yellow, and black as a flame or live embers. Other myths insisted the marks were a sort of divine calligraphy from major gods or minor saints. At least one tribe of American

Indians believed that the wispy lines were the entwined braids of lovers who had remained faithful until death. Of course, there is also the long tradition in Western literature of the painted face in the violet's flower. This may have begun with Martial, a witty poet of Nero's Rome, who complained that his suburban garden was so cramped that his violets dared not smile. Every century seems to have its flower lover who rediscovers the harlequin features on the heartsease *(V. tricolor)* and attempts to immortalize it in some sentimental note. Even Reginald Farrer, the great plant connoisseur and garden writer, couldn't resist remarking that the patterns on the heartsease "gave the flowers a most perky and intelligent expression."

Odor is a second cue that may determine how an insect positions itself on the flower. I suspect that the perfume glands (osmophores) of most *Viola* species are concealed primarily within the two fuzzy "beards" that decorate each of the lateral petals flanking the entrance to the floral tube. One observer has noted that when an insect thrusts its head into the opening, the beards hold the insect's body steady so that a load of pollen can be dumped on the forager's head and thorax while it probes for nectar.

Small wonder that Shakespeare seemed "critical" of the violet. Its fragrance and color seemed just a little too extravagant for such a small blossom:

> The forward violet thus did I chide:
> Sweet thief, whence didst thou steal they sweet that smells,
> If not from my love's breath? The purple pride
> Which on thy soft cheek for complexion dwells
> In my love's veins thou hast too grossly dyed.

When the flowers of some violets are carefully preserved, they retain their color and distinctive odor. Consequently, cultivated forms of the sweet violet *(V. odorata)* have had a long involvement with European cookery. During the fifteenth and sixteenth centuries the number of *Viola*-flavored dishes seems to have peaked. Flowers were simmered in honey, or soaked in oil, or made into a kind of marzipan with almonds and rose water, or even pickled in vinegar and sugar for "sallets." One of the ingredients of a popular hot soup from that period was made with a broth made of violet flowers, fennel seed, and savory leaves. Crystallized violet flowers are still regarded as an essential garnish for some classic Viennese pastries. Boxes of these sugared and dried flowers can be found, now and then, in German specialty shops in this country, but they are not very popular. My father remembers the Brooklyn candy stores of the twenties and thirties that imported crystallized violets, but he never cared much for the taste. "It was too much like eating perfume."

The range and intensity of floral color is yet another pollination strategy. The colors of violets of the same species can vary considerably from place to place. Just recently I was asked to explore the gardens of a new homeowner and list the plants. Barbara is extremely fond of wildflowers, and she greeted me with the happy news that her flowerbeds were overrun with three different "kinds" of violet. She pointed out a rare form with white petals hiding under the peonies. Then there were other clumps on the lawn displaying streaky-blue blossoms. Finally, the most common ones offered flowers that were a deep, royal purple. I hated to tell her that all three were the same blue woolly violet *(Viola sororia)*, the most common species here in Missouri. If you move down into the American South you may note that some of the "true blue" species also produce grayish forms and are often dubbed Confederate violets by local diehards.

Botanists are still just scratching the surface, but we do know that the genetic bases for these regional differences are rather complex. Jens Clausen, a Swedish pioneer in the study of gene distribution, tested color inheritance in European populations of heartsease or Johnny-jump-up *(V. tricolor)* back in the 1920s. No less than nine genes determine the ultimate intensities, hues, and patterns on the petals. Furthermore, these genes do not share equally in the production of color but form a hierarchy known as gene masking (epistasis).

There are three *A* genes, which control the amount of water-soluble pigment entering the flower. When all three genes are in a recessive state, most of the surface of the flower will appear chalky white to the human eye. The single L gene controls the splash of yellow on the spur and lateral petals. Two R genes work in conjunction to change red to violet. Finally there are three m genes, which suppress expression of velvety black borders on petal fringes. If the petals of a heartsease lack a black velvet border, it means that all three M genes are dominant. The encroachment of a black, furry border indicates that one or more of the genes are recessive. (See color plate 8.)

Every wild population of heartsease shows different frequencies for the range of color patterns. Since different insects may prefer different colors, the heartsease population is more likely to set a dependable crop of cross-pollinated seeds if it varies its communal "paint box" with each new generation. Should an important bee or fly species become uncommon in the habitat, there will still be other color patterns available to attract remaining insects or to exploit any new, nectar-thirsty visitors.

I did say, though, that some *Viola* species make a special commitment to producing seeds that goes far beyond the tricks of life span, movement, rewards, and attractants that I've discussed so far. During the first warm weeks of spring these violets offer their showy flowers for cross-pollination. As the season progresses, however, the plants send up increasingly shorter stalks and allocate fewer resources to the production of colored and scented petals. By late spring, or the first days of summer, the violet plant may still be manufacturing flowers, but these blooms resemble fat, green buds that never open. These are the cryptic, or cleistogamous, flowers, which can be found only by pushing aside the leaves and searching for these dwarfed stems toward the base of the plant. *Cleistogamy* means "closed marriage," and that is the fate of the cryptic flowers. The bud remains closed throughout the floral life span while two or more stamens remain wrapped around the receptive tip of the pistil. These cryptic flowers always self-pollinate to produce viable seeds, which, of course, are almost genetically identical to their single parent.

Gene masking in the heartease or wild pansy (Viola tricolor) *controlled by the three* M *genes. The flower in the
upper left-hand corner has no black borders, so it must be
dominant for all three* M *genes. The flower in the lower
right-hand corner has been tinted velvety black on each of
the five petals, so all three* m *genes are recessive.
Illustration by J. Meyers.*

There is sound adaptive logic to cleistogamy. Should the season for the cross-pollination of showy flowers be a poor one (due to late frosts, drought, pests, or an insufficient number of pollinators), there will nevertheless be a crop of seeds by summer. Violets are not the only plants that alternate showy and cryptic flowers. Similar strategies can be found in other local herbs, including the milkworts and gaywings *(Polygala)*, henbit mints *(Lamimum)*, bitter cress *(Cardamine)*, and wood sorrels *(Oxalis)*.

Moreover, because there is no cross-pollination or gene exchange, the cryptic flower's offspring are genetically programmed to live in the same environment as their parent. Sometimes this characteristic is very useful. The seeds of cryptic flowers grow into plants under the same conditions their parents did and produce both showy and cryptic blossoms, just as the offspring of cross-pollinations do.

Another way the shrewd violet increases its chances is in the way it disperses its seeds. Plants often use mechanisms to ensure that their seed is not merely dumped in a heap below the parent and eaten with relish by the small but hungry rodents and birds. Even unripened capsules are not safe from slugs and butterfly caterpillars, which devour them. Furthermore, seeds allowed to sprout too close together would compete with both their siblings and their parents for available resources.

To ensure the maximum number of survivors, most violet species broadcast their seeds over the widest possible area, and each species employs one or two methods to spread out offspring. Violets such as the smooth yellow *(V. eriocarpa)*, the common blue, and the bird's foot *(V. pedata)* produce a dry fruit called a capsule. As the capsule ripens, its walls dry out, and the capsule opens into sections called valves, which begin folding in on themselves. The strong valve edges cause the seeds to be expelled under pressure, catapulting them into the air.

Violets with such seed-expulsion mechanisms have stalks that are erect and usually higher than the leaves, so that there is no obstruction to the flight of the seeds, which may travel a yard or more. The smooth yellow violet *(V. eriocarpa)* holds the distance record in American species, shooting its seeds over 16 feet away from the parent plant.

Frankly, I never quite believed those distance records until I began growing plants of the ivy-leaf violet *(V. hederacea)* in Australia. One afternoon I noticed that a capsule on a favorite plant had opened and the brown seeds were lined up like soldiers on the margins of each valve. I thought that a friend of mine would want to see this, and as I started carrying the pot up a long flight of stairs, my footsteps must have shaken the plant so that the valves released all their seeds at the same moment. I heard a distinct *ping* as they richocheted off the ceiling, which was a full 6 feet over my head!

In contrast, the capsules of cryptic flowers put on a much poorer show. This is to be expected, since their seeds are best adapted for survival in the same "microhabitat" as the parent clumps. Consequently their valves shoot the seeds maybe only half the distance of the showy flower capsules. A couple of uncommon species in Kentucky and Tennessee may not shoot the seeds of their cryptic flowers at all. When the cryptic flower buds form on the rhizome, their stalks are so short that they do not emerge above the soil line. The young seeds start life preplanted underground!

To increase the chances of random dispersal even more, many violet species have seeds tipped with tiny oil glands called caruncles. Ants collect the seeds, carry them home to their nests, chew off the edible caruncles, and then dump the seeds in the colony's garbage heap, where they sprout.

Eleven Eurasian violets have dispensed with ballistics entirely and are completely dependent on ants. Violets such as the sweet violet *(V. odorata)* have flower stalks that collapse onto the ground when the capsule is ripe. The capsule's valves open gently and the seeds tumble out but don't stay piled up for long. Their caruncles are far longer and oilier than those species that only partially depend on ant dispersal. So adapted to ants is the sweet violet that the number of germinating seedlings increases slightly after the seed coats have been scraped by the mandibles of the ants.

Violets have a great tendency to form natural hybrids. That this occurs is not surprising. Different species may share the same

insect pollinators, and the flowering periods of many species over-
lap. In fact, when a stubborn winter delays the flowering of spring
herbs for several weeks, completely different species may flower
within a day of each other when good weather finally arrives. The
hybrid offspring of these unions provide a challenge to taxonomists.
As early as 1924 over eighty naturally occurring hybrids had been
recorded in the northeastern half of the United States. These natural mongrels are based on the repeated dalliances of only thirty
native species.

We know that these intermediate plants are true hybrids thanks
to the efforts of a New Englander named Ezra Brainerd, who
carried out breeding experiments and field observations for seventeen years. Mr. Brainerd must have really loved these wildflowers,
since he named his daughter *Viola.* And she was equally as enthusiastic as he, helping him make over sixty-one crosses between *Viola*
species and raising twenty sets of hybrids to maturity.

Hybrids are an important survival tactic of the genus *Viola.*
While they do not usually become new species—the first generation
of hybrids tends to show high levels of infertility, so hybrid lines
die out eventually if hybrids mate only with hybrids—they do
represent a source of new genes for the parent populations. Backcrossing of hybrid offspring with one of the two parent species is
a mechanism by which genes of one species may "filter" in to the
other species.

After several seasons of backcrossing, the parent species contain races with new characteristics, which enable the old species to
survive different environmental conditions and extend their ranges.
For example, hybrid offspring of a forest-dwelling violet and a
meadow violet, when backcrossed with either parent, will contrib-
ute new genes to the parent populations. As a result, future offspring of the forest violet may have some genes that enable them
to stand greater exposure to the sun, and meadow violet offspring
may receive genes that make them more tolerant of shade. The
variation of leaf form in such eastern American violets as the wood
violet *(V. palmata),* the coast violet *(V. brittoniana),* and the plains

violet *(V. viarum)* strongly suggests that each species has been enriched and developed by backcrossing.

Hybridization has its uses to the horticulturist as well. Where would our modern pansies be without it? Just as the European cowslip *(Primula vulgaris)* and the wild primrose *(P. veris)* were artificially crossed to give rise to the first polyanthus, the cultivated pansy is also a child of mixed heritage. During the early nineteenth century, cultivated breeds of *V. tricolor,* already a popular plant in English gardens, were crossed, recrossed, and backcrossed with three other species: the yellow pansy *(V. lutea),* the alpine violet *(V. altaica),* and the horned violet *(V. cornuta).* This process of selective breeding has resulted in such bedding flowers as the tufted pansies, violettas, and the large, flat, circular, monkey-faced "French" pansies we appreciate today. In fact, pansies seem to be getting bigger all the time, and that's a pity. What's so exciting about a flower the size of a tea saucer that flops in the mud and has lost most of the angular characteristics and sharply contrasting colors that made the wild species so appealing to the great poets and naturalists?

Let's discard the sentimental notions of timidity. Violets survive because they keep all their reproductive options open, whether they attempt to colonize sand dunes on an Irish beach or maintain small, yet enduring clumps in a farmer's woodlot in Vermont. They continue to serve both botanists and horticulturists as living laboratories because they demonstrate the principle of survival of the "reproductively fittest." It is no accident that yet another name for *V. tricolor* is love-in-idleness. Yes, this is the same little western flower whose milky juices caused the amorous goings-on in an Athenian forest one midsummer's night.

Chapter 11

In the Shadow of Forest and Glacier

What I want is light upon the whole of Nature—her methods, her laws, her results, her non-human ways.

—*John Burroughs*

PLAYING THE ROLE of the expatriate botanist for nine years taught me a simple truth: some plants live "without honor" in their own countries. I remember the teenager in the Salvadoran Orchid Society who would tolerate his diverse, native flora (see chapter 14) but never stopped pestering me about American sources who might sell him a lady's slipper orchid (see chapter 13). Then there's my Australian friend who lost his job as head gardener in a rural town, in part because he was always trying to use native shrubs in the local parks and was accused of not paying enough attention to the annual dahlia display.

One man's wayside flower is another man's cherished exotic. We Americans are just as myopic as any of these people I've just described. Would you be willing to spend over three dollars for a trout lily? That's what these common American herbs sell for in an Australian nursery catalog. Here in the United States the members of the genus *Erythronium* fall prey to a reverse snobbery. The Missouri Botanical Garden contains a lush, woodland garden that has been landscaped in the English style. Currently, though, this garden gem fails to exhibit a single trout lily!

No one has ever tried to monograph the genus *Erythronium* in its entirety. Consequently, there are twenty-three or thirty-two species distributed throughout the temperate zones of the Northern Hemisphere, depending upon which source you quote. Only one species is native to Europe, and its blood-purple or deep-lilac blossoms gave the genus its name, which derives from the Greek word for "red," *erythros*. However, in Canada and the United States, the species found along the Atlantic and Pacific coasts bear yellow flowers, and much further inland the dominant species are white with glossy blue or pinkish highlights.

Throughout Europe the common *E. dens-canis* is often called the dog's-tooth violet. The word *violet* refers to the flower's color and not to any resemblance with the plants of chapter 10. For the longest time, though, I didn't understand what made *Erythronium* species especially "toothy." William H. Rickett, former wildflower specialist at the New York Botanical Garden, explained that it's the underground, pointed, storage organ (technically known as a corm), which has the appearance of a smooth, white fang. Americans often refer to their native *Erythronium* species as dog's-tooth violets, much to the despair of generations of naturalists.

John Burroughs, a nineteenth-century nature writer and poet, fought against the use of the Continental name with true Yankee obstinacy. He suggested we use trout lily or fawn lily as substitutes. Other authors of wildflower books refer to the *Erythronium* species as adder's tongues, glacier lilies, or Easter yellow lilies. You may think it silly to list all these names, but I've found they actually help to describe the features of the plants and tell us quite a bit about their life cycle and distribution. American naturalists feel that common names can be pretty but should also be informative.

The deep-green leaves of *Erythronium* plants are often mottled with magenta or brown spots, and the floral face often wears dark freckles. This is reminiscent of the speckles on a trout's body or the camouflaging dapples on the back of a newborn fawn. Additionally, most *Erythronium* species flower sometime during the American spring and are regarded as harbingers of both the annual trout season and the birth cycle of our deer.

When *Erythronium* leaves begin to push through the soil's crust and last year's accumulation of autumn twigs and leaves, the new foliage is dark, curled, and pointed. This reminds some people of the forked tongue of an adder. Each mature plant also produces a single stalk tipped with one nodding flower that bears six long red stamens and their jiggling anthers, also resembling a snake's tongue.

The largest species of *Erythronium*, the aptly named *E. grandiflorum*, is common to the west coast of Canada, Washington, and Oregon. It is almost the only New World *Erythronium* to flower above the timberline on mountain slopes watered by melting glaciers, where it survives the longest, coldest winters and may not begin to flower until June. In effect, it follows the retreating slabs of ice.

The remaining species usually have flowered in time for Easter, but don't mistake them for perfect, miniature replicas of common garden lilies such as the madonna or the tiger. It is true that all of these plants belong to the same huge family, Liliaceae, and share some characteristics, but our garden lilies are in a separate genus, *Lilium*. Both genera grow a swollen storage organ that is really a modified underground stem. Both produce floral organs in multiples of three. I've already mentioned the six stamens, which both share, but they also have a single pistil with a three-chambered interior. We do not bother to distinguish between sepals and petals in either *Lilium* or *Erythronium* species. The six, brightly colored structures that ring the sexual organs are so similar in shape and position that we just call them tepals.

So much for the similarities. These basic structures show divergent trends in *Lilium* and *Erythronium*. The storage organ of a *Lilium* is a large bulb clothed in dense scales. *Lilium* flowers emerge at the tips of long leafy stems. The flowers hide their nectar glands within the three grooves on the triangular pistil.

Erythronium species, on the other hand, have small storage organs with tiny, transparent scales that don't completely clothe the white, toothy corm. The flower stem of a trout lily is free of leaves; its two leaves are flush with the ground. If the trout lily has nectar glands, they are placed at the base of each of the innermost tepals. It would seem that trout lilies have more in common with tulips (see chapter 7) than with tiger lilies.

The majority of *Erythronium* species in North America are successful colonizers of the forest floor. Therefore, like the wake-robins *(Trillium)*, bellworts *(Uvularia)*, Solomon's seal *(Polygonatum)*, and so many other American members of the Liliaceae, ecologists classify trout lilies as "spring ephemerals." They grow best in forests of maple, oak, beech, and hickory, which shed all their leaves every autumn and do not replace them completely until after the freezing weather ends and several weeks or months of spring may have elapsed. Sometime between the last March frost and the complete unfurling of the forest's new foliage, the forest floor experiences several "waves" of spring ephemerals.

Trout lilies must compress a season's worth of food making and reproduction into a couple of weeks, because when the forest canopy fills out, light levels on the ground drop dramatically. Insufficient light means a poorly photosynthesizing wildflower and blossoms that can't be seen by pollinating insects. Therefore, the growth of trout lilies must be rapid and precisely timed: leaves wither and pistils ripen with seed before the natural shrouding is completed. For the rest of the year the plants live on the nutrients stored in their corms. What is more ephemeral than flowers that lie down and die in darkness?

And although trout lilies may be dormant most of the year, let me emphasize that they are not merely an attractive garnish of our forests. Research shows that they are an intrinsic link of the cycling of both plant and animal nutrients in our woodlands.

As late snows melt and vernal showers soak the earth, running water leaches forest soils of the trace minerals vegetation requires to sustain itself. Trout lilies begin their vigorous growth just as this annual "fertilizer drain" is taking place. Their roots retrieve molecules of phosphorus that would otherwise be lost to the ground water. One ecologist thinks of trout lilies as living phosphorous sinks, with the mineral transferred to their green, uncurling leaves to be used in the construction of enzymes and fat compounds. When the leaves wither, the phosphorus is returned to the rich forest litter by bacteria and other microorganisms so that other trees and shrubs may use the precious molecules during their later periods of active growth.

Trout lilies are insect pollinated. There is a very close relationship between the plants and several species of bees and large flies, which have either overwintered in a cocoonlike (pupal) stage or hibernated as adults. When they emerge in April and May, the nectar and pollen of trout lilies sustains them until other wild-

flowers start to bloom. Trout lily pollen appears to be very important to some mining bees, which dig their burrows in the earth.
Females live solitary lives, although two "working mothers" may
share the same burrow, and, in fact, burrows of different bees
may often intertwine and interconnect to form "bee condos." The
best-known miner bee is *Andrena erythronii,* whose name makes
it clear that it is very dependent on the spring crop of *Erythronium* flowers. The female *A. erythronii* lays about eight eggs and
also lays in a store of trout lily pollen for when her hungry larvae
hatch.

Pregnant queen bumblebees (technically known as gynes) are
another important group of foragers. A queen bumblebee waking
in the spring is an orphan—her queen mother, drone husband, and
worker sisters do not survive winters in this latitude. Each autumn,
next season's gynes leave the family nest and hibernate in an old
log or under boards until they emerge during the early, warm days
of spring.

Compared with the queen honeybee, who is never without a
retinue of neuter workers, the queen bumblebee works indepen-
dently to start a new colony. She finds a place to nest (the abandoned burrows of field mice and other rodents are often utilized)
and then raises the first generation of workers all by herself. The
sight of these large, fat bees each spring terrifies many people, who,
when approached by the hovering gynes, go into hysterics and grab
the first blunt object they can get their hands on. While I am
sympathetic to those with severe allergies, I want to say that the
gyne rarely stings unless she's grabbed or sat upon like Ferdinand
the bull. When she flies about, she is either looking for a place to
nest or merely attempting to memorize local landmarks near her
new nest while she's out shopping for pollen.

It's interesting to watch a bumblebee queen as she methodically
visits trout lily blossoms. She dangles upside down, like Quasimodo,
while gathering pollen or sucking nectar. One pair of her legs clings
to the tepals while a second pair scrapes pollen from the red,
yellow, or white anthers. The last two legs temporarily store the
pollen: trout lily grains are packed into a smooth hollow located on

1. *During the dry season, the trees of the lowland forests of Central America exhibit different flowering patterns. The limbs of this cortes (*Tabebuia sp.*) offer a "Big Bang" strategy, as the majority of flower buds open at the same time. Although this individual offers lots of nectar to local insects and hummingbirds, it remains in flower for less than a week.*

2. *Comparatively few of the tree species of the lowland forests retain their leaves throughout the months when no rain falls. Forest colors are subject to dramatic changes during the dry season when trees flower following leaf drop. These yellows have been produced primarily by tropical members of the trumpet vine family (Bignoniaceae).*

3. *Sensitive brier or cat's-claw* (Schrankia nuttallii) *is a common wildflower of late spring on the tall grass prairie. It is a mimosoid legume, and the feathery leaves do close when touched. The flowering heads have a mild but unusual odor reminiscent of raw green beans and caraway seeds. Although the purple florets that make up each head lack nectar, they are still visited by more than 20 species of bees collecting pollen for their offspring.*

5. *The hairy body of* Amphicoma vulpes *rubs against the blackish stamens of a scarlet anemone* (Anemone coronaria) *as the beetle searches for pollen. Although this flower exhibits the common salver/ bowl form and may be visited by almost any hungry insect, the* Amphicoma *beetles of the Middle East remain the major vehicles of cross-pollination.*

4. Opposite: *Tall grass prairie in late May. The tillers of grass reach no higher than your ankles at this time of year so the habitat is a carpet of the shorter species of wildflowers, including the greenish pom-poms of* Asclepias viridis *(a large milkweed) and purple drifts of prairie turnip* (Psoralea exculenta), *which is really a member of the bean family. Shrubs and small trees are limited to rocky outcrops and stream banks.*

6. *Even an immature leaf of* Victoria amazonica *can support the weight of an equally immature member of the Yagua Indian tribe. The flower of this plant rivals the size of the baby's head. This will be the last day of life for the white blossom, which will turn a deep burgundy by late afternoon.*

7. *Mature leaves of the hybrid* Victoria amazonica *X* V. cruziana *float on the surface of an indoor pool at Longwood Gardens.* Victoria *leaves develop a stiff, erect rim unlike the leaves of most other water lilies. However, this rim has two notches, one on either side of the leaf, to permit the easy run off of rainwater following a storm.*

8. The bird's-foot violet (Viola pedata) *is found in open woodlands and rocky fields throughout much of the eastern half of the United States from March through May. Unlike so many other violet species, the petals of the bird's-foot lack a beard and the plants do not produce cryptic (cleistogamous) flowers after their showy blossoms are spent. Most bird's-foot violets bear flowers with five lilac-colored petals, but this individual shows a natural trend toward melanism.*

9. Cypripedium *is atypical of most orchid genera, as it shows its greatest diversity in the cooler regions of the northern hemisphere. The yellow moccasin of yellow lady's slipper* (C. calceolus) *enjoys a natural distribution encompassing parts of North America, Europe, and temperate Asia. Although Americans often associate this species with forest glades in late spring and early summer, it has evolved localized varieties adapted to bogs, limestone barrens, and prairie edges.*

10. European horticulturists attempted to grow the Cymbidium *species of tropical Asia as early as the 1830's without much success. Although this* Cymbidium tigrinum *comes from the mountains of Burma, it requires cool night temperatures to thrive and flower—not the monotonous steambath produced in stovehouses and early greenhouses. Unaltered by hybridization and selective breeding, this plant maintains a sinuously erect, few-flowered stalk. Each flower offers a long, curved, freckled labellum in contrast to its narrow, olive-colored sepals and lateral petals.*

11. In contrast, this Cymbidium "Mighty Oloroso" typifies the breeder's attempt to suppress natural floral features. The labellum has been shortened and the remaining petals and sepals have been bred for thickness and width to give each blossom a flattened, circular appearance. So many large, heavy flowers now crowd a single branch that mature stalks must be propped up with stakes.

12. The butterfly orchid (Oncidium papilio) *produces one of the largest blossoms recorded within the more than 400 species that comprise the genus* Oncidium. Oncidium *species are native throughout the tropical Americas, and four species grow as far north as Florida. Known as spray orchids, dancing dolls, mule ears, and* chorizo con huevos (sausage with eggs), *these species occupy a wide variety of habitats, but most establish themselves on tree limbs or boulders. This flower displays the glossy yellow-brown splotches and extravagant central "lumps"* (Oncidium *means little tumors) common to the genus.*

each tibia of the hind legs. This depression is encircled by a fringe of inward-curving hairs to form the pollen basket of social bees called the corbicula. Flower nectar gives the queen both a source of energy and a "pollen mortar," which helps keep the grains packed down within each basket. Some queen bumblebees carry so much trout lily pollen that the contents of their baskets grow round and smooth until they are almost as large as aspirin tablets.

Pollen that has been "sugared" with nectar is fed to the queen's brood. However, we know that at least one white trout lily *(E. albidum)* does not offer any nectar. Apparently the grains are so rich in fat and amino acids that bumblebee queens continue to visit the flowers faithfully without ever receiving a liquid reward. Under these circumstances, the gyne becomes a selective shopper, visiting a few *E. albidum* for pollen and then leaving to take a nectar break at other early, spring flowers such as Dutchman's breeches *(Dicentra)*. If she didn't sip nectar, she might run out of energy and not have enough strength to return to her own nest.

The bumblebee larvae fed on trout lily pollen during the spring mature to pollinate clover, alfalfa, and other early crops. The trout lilies form an early but essential link in a chain of different flowers, separated in time, that utilize the same pollinator. When trout lilies fade, the bumblebees pollinate the trailing arbutus *(Epigaea)* and then possibly mayapple *(Podophyllum)*. When these ephemerals lapse into dormancy, the bumblebees move on to the marsh and meadow plants. On it goes until the autumnal frosts, when a new generation of bumblebee gynes go into "cold storage" to await the floral merry-go-round of next year.

The fertilized pistils of most trout lilies ripen into capsules as trees' new leaves begin to shut out the sunlight. When ripe, the stalk collapses and the valves of the capsule open, spilling the seeds onto the forest litter in a little pile. Here the trout lily makes one last contribution to the nutritional needs of forest insects. Like some of the true violets of chapter 10, the seeds of trout lilies also are tipped by soft, often meaty caruncles. Observations by Douglas Schemske in Illinois suggest that American trout lily seeds are dispersed by hungry crickets and ground beetles that drag them

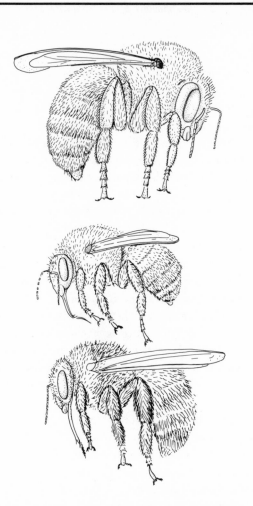

Each spring the young bumblebee queen (top) collects pollen to feed to her first generation of workers (middle). The first workers tend to be rather small compared with the queen. These "runts" care for a larger generation of sisters (bottom) although no sister is ever quite as big as her queen mother. Illustration by J. Meyers.

away and gnaw on the caruncles. This usually happens during the evening.

In Japan, *E. japonicum,* seems to have more efficient seed dispersers. Common ants *(Aphaenogaster famelica)* visit the little seed piles and carry them all away within an hour and a half. They leave scent trails between the seed piles and their anthills so that they can return to their food source repeatedly. It's interesting to note that the seeds of *E. japonicum* begin to sprout almost as soon as they leave their capsule. In contrast, the slowpoke seeds of the eastern American trout lily *(E. americanum)* do not begin to germinate until early autumn, nearly five months after dispersal, and the seedling does not leave its husk until after the big thaw the following year!

It may seem rather paradoxical that many *Erythronium* species use very little of the food reserves in their corms to make seeds. Even nineteenth-century botanists like John Torrey noticed that while the pistil of *E. americanum* could contain over a hundred immature, unfertilized seeds (ovules), most of them would fail to develop—a single capsule rarely contained more than six to eight good seeds.

Over the past decade this discrepancy between ovule and seed number has distressed a number of population biologists who work on woodland flowers. We expect flowers to make a surfeit of pollen, since so much of it is eaten by pollinators or is lost en route to the pistil tip. Wasted eggs, enclosed in neglected seed bodies are another matter, however, and have been a source of speculation for years.

I call one of the explanations for this phenomenon the "you just can't get decent servants these days" story. Some scientists blame the pollinators for the unfertilized ovules, claiming that the flowers of trout lilies are poorly patronized. This hypothesis has been largely discredited in the past couple of years by fieldworkers who spend long hours watching the pollen-harvesting techniques of the insects, following them from flower to flower and then using a hand lens to count the number of grains bees leave on a pistil tip. Mining bees and bumblebees are faithful, in their fashion, and they certainly leave more pollen grains on a stigma than the minuscule seedset would suggest.

I call a second explanation the "wise mother" tale. In this scenario, the pistil makes excess ovules as a kind of damage insurance. If some eggs meet with the wrong kind of sperm, the parent has the chemical option of aborting them. Only the most physically and genetically fit seeds survive. My complaint with this more intriguing hypothesis is that scientists have not defined the nasties that may attack a pistil and ruin some seeds but leave others unscathed. No breeding experiments have been completed to see if some plants actually produce pollen that is deemed radically inappropriate by their neighbors.

In my opinion, the unused ovules should be treated as vestigial organs like the human appendix. Trout lilies probably evolved from sexually reproducing plants that utilized most of these tiny organs to make seeds, but it is possible that the role of ovules has declined in importance as trout lily populations have become more dependent on another mode of reproduction. In North America most trout lily species are colonial, with colonies resulting from asexual reproduction. In a few species new corms bud off the bodies of old corms and form offsets. However, most trout lilies make droppers.

A dropper is a tubular, fleshy bud that grows from the base of an old corm. The dropper's stem (or stolon) penetrates deep into the soil until its tip begins to swell, and then it grows into a new corm. Wild trout lilies are so difficult to dig up intact because just as you think you've reached the storage organ, the plant snaps off on your spade while the corm remains safely hidden under another layer of earth.

The Japanese trout lily and the glacier lily do not often reproduce through clones, and their mature corms tend to make budding offsets only after they are injured. However, between 30 and 90 percent of the ovules in their pistils mature to form good seed after cross-pollination. Compare this with the eastern American species such as *E. americanum* or *E. umbilicatum,* both are extravagant cloners, and only 10 percent of their ovules survive to seedhood. I rest my case.

A colony of trout lilies usually consists of one or two two-leaved plants and up to ten one-leaf plants. The one-leafers can be clones of either a two-leaf or a one-leaf plant. Since trout lilies always produce two leaves when they are ready to flower, there is considerable debate regarding the fate of the clones.

The more conservative opinion expressed by the New York botanists Henry A. Gleason and Arthur Cronquist accepted the theories of earlier naturalists who contended that one-leaf clones were doomed to a lifetime of sterility. Dr. Cronquist once wrote to me to suggest that if a one-leaf plant eventually produced two leaves and a flower, a colony would have more and more flowering plants every year. And if this is so, why, then, are there so few flowering plants in a colony? Cronquist held that to produce a flower a plant has to grow from a seed rather than a dropper.

William Rickett and the Michigan botanist H. V. Smith offered a second theory. Since the growth cycle of trout lilies is so brief, it may take some time before a one-leaf can become a two-leaf. During that long wait until adulthood, the fertile parent plant will be eliminated by environmental pressures or die of natural causes. The "flower torch" will be passed to the oldest surviving clone, thus keeping the sexual cycle moving.

What do I think? Well, we still need more work on this topic, but there seems to be some evidence to support Rickett and Smith. Trout lilies do take a long time to reach reproductive maturity. A seedling of the Japanese species waits eight years until it is big enough to flower. I've noticed that most colonies of trout lilies on the East Coast follow the few two-leaf/many one-leaf scheme. However, every now and then one finds a woodlot or glade where conditions seem to be perfect for a floral explosion.

When I was doing my master's degree in Brockport, New York, I studied *E. americanum* in a little reserve called Swartout's Woods. By April the woodland ground had become a living carpet of trout lily leaves. There were plenty of sterile clones, of course, but when I checked the moist, humus-rich dells (created by the thick, old roots of trees) I found as many as twenty flowering plants packed together. It seems that there are certain infrequent circumstances

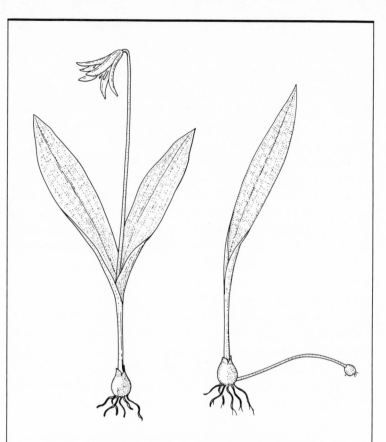

The two- and one-leaf forms of a trout lily or adder's tongue (Erythronium americanum). When there are two leaves, the plant produces a single flower each spring (but it can also form a dropper). One-leaf plants can reproduce only by means of a dropper. Illustrations by J. Meyers.

under which clones can become two-leaved, flowering plants. Perhaps overcrowding is the clue. Douglas Schemske told me he's observed the same massive flowering in populations of *E. albidum* in Illinois.

One thing is demonstrable. North American trout lilies are geared to self-clone. Robert Muller, at the University of Lexington, Kentucky, calculated that an *E. albidum* plant commits about 58 percent of its energy reserves to the production of a single clone, while one seed uses only 3 percent of the plant's resources.

Why continue to make seeds at all? Why don't all trout lilies cash in their seeds and stick to dependable droppers? One species has done this, in fact, and it provides a remarkable example of the perils of abandoning gene flow.

Jo Ann Banks examined the home life of *E. propullans* in Minnesota. This species has rather tiny flowers that produce a lot of dud pollen. Mining bees visit the flowers, but they never seem to set seed. Jo Ann even tried hand-pollinating the flowers and had no luck. The plants appear to increase exclusively by droppers. Since Jo Ann could not get any of the *E. propullans* plants to set seed no matter how pollen was exchanged between different colonies, she wondered if the entire species was just one continuous clone that was slowly being divided by natural mortality and invasive vegetation.

It's interesting to note that *E. propullans* is a restricted species, known in only two counties in southeastern Minnesota. Without seeds this species cannot extend its range very far or very fast. Without cross-pollination this species cannot filter new, beneficial genes through its remaining population. Since each colony is rather similar, if not identical, to its neighbor, it means that if one colony suffers the effects of a natural disaster, they all will. So, seeds may be the best tickets in the lottery of life. Most *Erythronium* species continue to conquer the forests with their clones, but they also take care to feed the bees, ants, and crickets as a reward for long-term services well rendered.

On the whole, intelligent North American gardeners feel that the proper place for a trout lily is its natural environment. It's considered bad form to dig them up. When dams and industry destroy forests, modern conservationists look at shrinking habitats and suggest some attempt at artificial culture. This will become more important if we wish to protect the less common species of trout lily.

Chapter 12

Hanging by Their Heels

It has been my general experience that pessimism about the adaptive significance of a trait is strongly correlated with ignorance of the natural history of the organism.

—*Daniel H. Janzen*

A LARGE NUMBER of popular and interesting groups of plants have been named after obscure and uninteresting people. This is a tradition from the great age of botanical exploration, when plant scientists endured great risk, expense, and loss in their quests for the rare, the beautiful, and the new. Field collectors succumbed to disease, aggressive animals, and unsympathetic natives. Ships sank en route to Europe, destroying every last shred of evidence that a botanical exploration had ever taken place. Carefully packed specimens were often devoured by stowaway rats and roaches. To name an entire family, or genus, of plants after one man seems terribly extravagant, but this is the way scientists thanked both their peers and their patient employers for their moral and financial support. Frequently, too, plants were christened after individuals who played no part at all in their discovery or subsequent care. We can laugh at this floral tribute, but it still goes on, albeit at a more restrained level. *Dressleria*, for example, is a recently named genus of bizarre and beautiful orchids. It's not really fair that such stunning flowers were christened after a rather homely botanist whose Latinized name sounds like the name of a clothing store in a suburban mall.

The flora of South America have often fallen victim to this form of botanical backscratching. Transport to Europe was arduous and the survival rate of tropical material was poor. Glasshouse culture was not practiced until well into the second half of the last century (see chapter 15), so live cargo would often perish even if it managed to make it across the Atlantic. The few survivors were considered precious and were, therefore, named for minor scientists or those wealthy dilettantes who sponsored the field trips. Many members of the pineapple family (Bromeliaceae) were named after those gentlemen who pioneered the young science of botany. Bromeliad genera with names such as *Billbergia, Guzmania, Vresia, Hechtia, Hohenbergia*, and so forth all belong to the same exclusive men's club.

The father of the modern system of biological nomenclature, Carolus Linnaeus, named the genus *Tillandsia* after the Swedish physician Elias Tillands, never realizing that he had conferred a great honor on his friend. Tillandsias are now regarded as the most successful and widespread of the wild bromeliads. They are found from the southeastern section of the United States down through the tail of the South American continent and all through the tropical isles of the western Atlantic. It is the largest genus in the Bromeliaceae, with over four hundred individual species described, not including the many additional subspecies, natural varieties, and hybrids.

Most North Americans are familiar with one famous tillandsia, if only because it has made an appearance in almost every classic Hollywood film set in the Old South. Few people realize, though, that Spanish moss *(Tillandsia usneoides)* is a bromeliad, let alone a flowering plant. It has the widest distribution of any *Tillandsia* species throughout the Americas. In the United States odd collections of it have been made only fifty miles or so south of New York City. It's all over the Caribbean and as far south as Chile and Argentina. This plant forms dense, gauzelike masses consisting of hundreds or thousands of individual plants. Each plant grows no more than three leaves before it flowers and dies. At one time an important cottage industry in the United States was milling the tiny plants and using them as a stuffing for upholstery. Today, Spanish moss is still regularly employed by horticulturists as a natural sunscreen in greenhouses and as a living humidifier to protect more delicate plants from desiccation. There are several more *Tillandsia* species distributed as far north as southern Florida. They are larger plants more commonly known as wild "pines" (short for pineapples), although they never produce succulent, edible fruit.

New species of *Tillandsia* are still being discovered today in poorly explored territories of Mexico, Panama, Brazil, and the Guyana highlands. In their native habitats tillandsias seem to grow in almost any substrate except rich, moist soils. Some find a convenient niche between boulders on a cliff or hug the walls of rocky escarpments. I have found tillandsias in gravel-encrusted depressions made by the hot-water fumaroles along the slopes of Central American volcanoes. The little bromeliads did not seem to be bothered by the frequent steambaths. Like most bromeliads, though, the majority of *Tillandsia* species grow on the branches or trunks of trees. They are true epiphytes, merely using bark and twigs as landing platforms. They do not actively parasitize their hosts like the mistletoes of chapter 3.

Some tillandsias have even recolonized spaces that man has covered with his cities. Throughout Latin America an odd but common sight are such plants as *T. schiedeana* and *T. recurvata* dangling from telephone wires. A few countries employ urban workers to keep the poles and cables clean of tillandsias and other bromeliads. Without this maintenance the poles could collapse and the wires snap under all the extra weight. Of course you needn't go to the tropics to see urban bromeliads. The shady live oaks of New Orleans and many other trees in Gulf towns and cities are often festooned with Spanish moss and other wild pines.

Tillandsias establish themselves on these perches by hit-or-miss tactics. The dry fruits open spontaneously in warm, dry weather, and the seeds are carried off by the wind. Each seed is tipped with a prominent cluster of plumelike hairs, known as a comus, which acts like a parachute. When the comus hairs brush against a rough surface, such as a bit of bark, they become entangled, and the seed ceases its wanderings. On tropical savannas, where trees are few and heavily laden with epiphytes, I've seen branches so coated with white down they look like the victims of a violent pillow fight.

The comus holds the *Tillandsia* seed in place until it germinates. When the seedling bursts its coat, it fastens itself to its host with tiny rootlets that serve as grappling hooks. As the plant's leaves grow larger and heavier, the roots may grow a little bit larger to lash it down more firmly. This period of establishment, when the seedling clasps its host and produces its first functional leaves, is the most critical phase in its life cycle. Most plantlets die in the attempt.

Field experiments conducted in Florida using *Tillandsia circin-*

nata, a native wild pine, show how common mortality is for seed-lings: six thousand seeds were artificially "planted" on the branches of sixteen trees in five different habitats. The comus of each seed was carefully removed and then the naked seed was stuck to the bark using Elmer's glue. When the trees were visited a year later, only 185 "babies" were counted—less than 4 percent of the original crop. Only five of the sixteen trees offered an adequate cradle. Seedlings were more likely to survive if they were planted on a native cypress *(Taxodium)* or a mangrove *(Avicennia).* Apparently the bark of only a few tree species produces an appropriate microgarden for the seeds of *Tillandsia* species. Most of the annual *Tillandsia* seed crop is wasted, since comparatively few seed darts hit the right targets.

Dissect the root of a fully grown *Tillandsia* and you will find almost no living tissue. It is only a piece of rope. If this organ has become so modified that it's incapable of absorbing water, just how do these plants stay alive? Most of the answers derive, in large part, from the meticulous field and lab research performed by Dr. David H. Benzing, of Oberlin College, Ohio, and his students and colleagues.

Many members of the pineapple family are known as tank bromeliads. Their leaves are flat, wide, and trough-shaped and interlock at their bases to form a natural crater. Rainwater and heavy fogs spill in and fill the craters. The water is absorbed directly through the leaves by osmosis, completely bypassing the roots. Although a few species of tillandsias, such as *T. micrantha, T. rubra,* and *T. guatemalensis,* utilize these trough-shaped, water-collecting leaves, the majority of *Tillandsia* species have leaves that are narrow and thin. Some are rigid and needlelike, whereas others resemble twisted pieces of wire. The leaf bases do not form a crater or vase. In fact, the bases of some leaves are rather thick and bulbous. Plants with such leaves often hang upside down and are often known as air plants, or atmospheric bromeliads, and the way they capture and retain moisture is quite different.

The leaves of most atmospheric tillandsias have a gray, scurfy surface. Take a look at a leaf under the microscope and this "psori-

Tillandsia caput-medusae *(left) usually hangs upside down by its roots. Ants establish their nests in the inflated base of the stem. The wide leaves of* Tillandsia grandis *(right) form an interlocking vase so that rain and fog collect in a natural tank, which offers a damp home to sowbugs and salamanders. Illustrations by J. Meyers.*

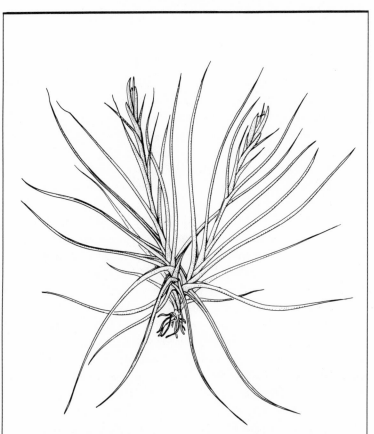

Tillandsia schiedeana *is a true atmospheric bromeliad, receiving water and dissolved minerals only through the scurfy scales on its narrow leaves.*
Illustration by J. Meyers.

asis" is revealed as a covering of specialized parasol hairs known as peltate trichomes. Each hair consists of a single stalk, a short chain of living cells, embedded in a cavity on the plant's skin. The stalk is tipped with a wide, beautiful shield that collects splashing raindrops or fog. The water molecules then enter the body of the leaf through the short stalk.

The shield is quite an elaborate structure featuring three different kinds of cells that are dead and hollow at functional maturity. The central disk of each shield is made of four cells, which have a thick upper surface but thin and permeable undersurface. The central disk is encircled by one or two rows of ring cells, which are then surrounded by a pretty fringe of long, thin-walled wing cells. The wing cells form the margin of the shield.

The fringe of wing cells curls upward on hot, dry days and forms a tiny cup that waits patiently for the arrival of moisture. When a raindrop strikes the cup, water is drawn into the shield through the thin, permeable walls of the hollow ring cells. This absorption occurs in less than a second. As the shield fills with water, the flexible wing cells collapse, becoming parallel to the surface of the *Tillandsia* leaf. The water that collects in the ring cells is transferred to the larger cavities of the disk cells, where it passes into the live stalk cells by osmosis and is then transferred to the base of the stalk and into the thirsty tissues of the leaf.

These hairs are such efficient water collectors that atmospheric bromeliads can subsist on the dampness and humidity of the air. The leaves hoard water in their storage tissues, so that many atmospheric species can survive long, severe heat waves. This is why certain *Tillandsia* species crowd the limbs of trees that make up the tropical forests, which may experience a dry season of up to six months. Tank tillandsias, in contrast, tend to be more common at higher altitudes, where dry seasons are shorter and the plant's living vase receives one or more baths of mist from low clouds almost every day.

While the accumulation of water presents little problem to most tillandsias, the acquisition of basic minerals can be quite difficult.

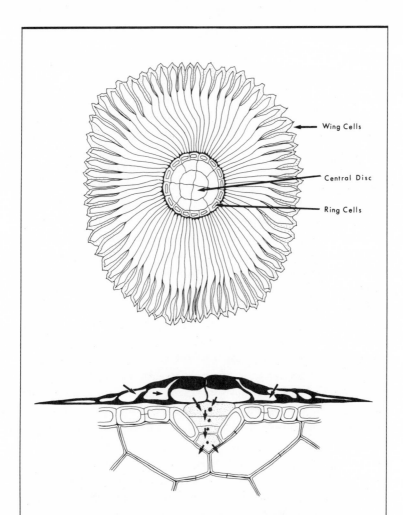

Wing Cells

Central Disc

Ring Cells

Overview of the scale trichome shield of Tillandsia circin-
nata. *All of the cells comprising the shield are dead and
empty at maturity (top). The flattened shield absorbs water,
which passes through the hollow cell chambers into the
living cells of the stalk (bottom). Once water passes into the
lowest stalk, or foot cell, it moves into the storage cells of
the leaf. Illustrations by J. Meyers.*

Since *Tillandsia* species do not commonly grow in soil, and their roots do not penetrate rocks or steal nutrients from the sap of their host trees, some critical elements are always in short supply. Sometimes a plant must enter a relationship with other tropical creatures that can lead to mutual benefits.

Tank tillandsias, for example, are living aquaria. Their craters offer a natural habitat to many arboreal insects, crustaceans, frogs, salamanders, and fungi, which must pass part of their life cycle in water. The decaying corpses and waste products of such organisms are absorbed by the leaves as a source of nutrients. On the island of Trinidad a few *Tillandsia* species and other bromeliad tanks create a special health problem. The warm water in the tanks is a perfect "nursery" for the wriggling larvae of the *Anopheles* mosquito, which transmit malaria as adults.

Depending upon the species, atmospheric tillandsias employ one or two tricks to accumulate minerals. Both mineral ions and amino acids pass through the cells of the peltate trichomes, provided they are diluted in water. Rain washes dust and debris from the surfaces of rocks and host trees, leaching out trace amounts of salts and simple organic compounds. As water runs down the branches, it spills onto the leaves of the tillandsias, and the modified hairs absorb a "leachate" that is higher in calcium, nitrogen, and phosphorus than plain raindrops that have fallen directly from the sky. Species that grow near the coastline absorb sodium and magnesium from the ocean's spray. Minerals and other nutrients move down the chain of cells in each hair stalk by active transport.

However, it has become increasingly apparent over the past forty-five years that some atmospheric tillandsias derive most of their nitrogen from the debris of ant colonies. Many tropical species of ants regularly invade the compact, bulbous bases and protect their plant home against all other invaders. I learned this to my regret when I was collecting plants on an old lava field in El Salvador, and I reached for a fat specimen of *Tillandsia caput-medusae*, but disturbed a vicious, stinging colony instead. These plants are technically classified as myrmecophytic tillandsias, plants that offer shelter, food, or both to ants. The ants are wonderful guards but sloppy housekeepers. Trash and dead ants may remain inside the living chambers until they are broken down by bacteria and absorbed by the tillandsia. Dr. Benzing proved this by placing minute amounts of radioactive calcium in leaf bases and then tracing the movement of the element as the plant transferred it to the coiled tips or reproductive structures.

T. caput-medusae is one of the most common and widely distributed of the myrmecophytes. With its inverted, globular base and its slender, serpentine leaf tips it really does resemble the head of the dead gorgon carried by the hero Perseus.

While tillandsias can subsist on levels of mineral nourishment low enough to starve other plants, their growth rate is slow. Even rather small *Tillandsia* species may take at least five years to produce flowers and seeds. One scientific journal described a large *Tillandsia* species collected in Mexico that languished in a botanical garden in Italy for nearly a decade before it put out a flowering stalk. Since sex requires such a great investment of resources for tillandsias, the plant that successfully concludes its reproductive cycle often dies soon after its seedpod opens. *Tillandsia* colonies are often a mixture of plants from several generations of seed and natural clones that began life as asexual buds developing at the base of an old plant as it started to flower. The original plant may die, but half of its genes live on in the seeds derived from its cross-pollinated flowers, and all of its genes survive in the young, perfect replicas clustered like piglets around a sow. Theoretically a *Tillandsia* colony should grow larger and larger, like a coral reef. More often than not, however, a storm blows it down or the branch on which it is perched breaks under the weight.

Reproduction is so slow a process for tillandsias that a plant may remain in flower for more than half a year. The blossoms are tubular and range in color from white to pink and from blue to velvety purple. Individual flowers are rather short-lived and rarely last more than a day or two. However, as each flowering spike often conceals dozens of buds, a new flower may open each morning over a period of months. Of course, this means that *Tillandsia* spikes will rarely offer more than one or two open flowers at a time.

I doubt that people would prize tillandsias very much if all they had to offer was a persistent dribble of pastel flowers. In most species, though, each flower bud is enclosed by a brightly colored bract that is usually larger than the mature flower and much more spectacular. They range in color from glossy greens to hot reds combined with vivid yellow stripes. By arranging pale flowers against the intense hues of the bracts, the whole spike of the tillandsia becomes a brilliant, long-lived billboard that stands out against the canopy of the host tree, attracting the local guild of pollinators.

Tropical hummingbirds appear to be the most important pollinators of *Tillandsia.* They may fly to the same spike day after day, taking nectar from fresh tubes as old flowers are replaced. Naturally, one or two flowers will not offer enough sugar to satisfy a single bird, but this encourages hummingbirds to visit every flower on a spike and then visit the spikes of other plants in the population. Since the bracts live on until the death of the parent plant, hummingbirds can easily locate the same plant over its flowering season. A hummingbird that traplines the flowers of coral trees (see chapter 1) can probably familiarize himself with the long-flowering *Tillandsia* species in his territory.

Not all animals that find the floral tubes are dependable pollinators. The so-called stingless honeybees *(Trigona)* are some of the most common tropical bees and are equipped with powerful jaws. These small insects lack tongues long enough to reach the reward at the end of the tube. I've watched them gnaw holes at the base of *Tillandsia* flowers and steal the nectar.

It's the clusters of bright spikes that give tillandsias their common names in Mexico and Central America. People call them *gallitos,* which is Spanish for "little rooster," since the clusters look like a red cock's comb. When bracts are yellowish, rather flat, and arranged in branched spikes, the plants are called *pie de gallo,* or "rooster foot." Species with the largest and most brightly colored spikes are also known as *guacamayas,* or "macaws," after the big parrots with streamer tails. There are many more variations on such names in Mesoamerica, where the oldest names derived from the original Aztec and Mayan tongues. In fact, this is one of the few areas where Hispanic names for common epiphytes are more imaginative than English names.

Man, the thoughtful enemy, has exploited *Tillandsia* species for as long as he has lived in the New World. Spanish botanists, who arrived in Peru after the conquest, noted that the Incas had many uses for *Tillandsia* and other bromeliads. They were draped over unripened fruit and used to cushion ceramic ware through long journeys. Tillandsias were useful construction materials, since the chopped-up leaves could be mixed with mud and cements to form a binder. Species such as *T. maxima* and *T. rubella* are still regarded as wild vegetables. Their young, cooked leaves are considered to be as tasty as palm hearts. Old plants are still knocked off trees and shrubs as an emergency browse for sheep.

Tillandsia species also found their way into a series of folk remedies. They were supposed to act as an invigorating stimulant when added to a hot bath. However, the most gruesome medicine must have been made when they were pounded into a paste and applied directly to hemorrhoids! Imagine the sensation of all those specialized hairs, with their wide, winged shields, on severely irritated flesh.

Tillandsias give us a good example of how some plants "change religion" when a native culture succumbs to conquering invaders. The Aztecs of the Yucatan peninsula decorated their temples with tillandsias, especially Spanish moss. The arrival of the conquistadors meant that Tillandsia crafts could be absorbed by a new pantheon but it's often hard to tell whose festivals received more honor. Much later, in fact, as late as the 1950s, men decorated their hats with garlands of it and nipple fruits *(Solanum mamosum)* for annual pilgrimages. They may have been on their way to visit an important church, but such unusual hat decorations are believed to derive from ancient fertility rites.

Tillandsias are still part of a charming yet environmentally destructive custom at Christmastime. Go to any southern Mexican or Central American open-air market in December and you will find tillandsias sold with lichen-encrusted branches and sheets of true moss. They are used to make miniature landscapes for the nativity crèches. Some very pretty effects are achieved with the Three Kings plodding over a green carpet through a forest of tillandsias.

Despite the great number of bromeliad societies scattered over the world, I can't help but think that we've all become a little blasé about tillandsias and have begun to take them for granted as rather vulgar inhabitants of tropical greenhouses. Despite the great number of species, *Curtis Botanical Magazine* (the "gourmet" catalog of the Royal Horticultural Society) has not featured a *Tillandsia* species in over thirty-five years. I gave up searching when I got back as far as 1949.

— Part Four —

Those Unnatural Flowers

Chapter 13

Lady's Wear

The spring is not perfect to the lover of May
who does not find the Pink Lady's-Slipper.

—*William Hamilton Gibson*

IN 1753 CAROLUS Linnaeus assigned the formal name *Cypripedium* to the genus of slipper orchids. *Cypripedium,* surely one of the most enchanting of all scientific names for plants, alludes to the foot (or shoe) of "the Cyprian One." That, of course, is the goddess Venus, who rose from the sea off Cyprus and returns each year to bathe and restore her beauty and her maidenhead. Today these exquisite terrestrial orchids have become victims of their own extraordinary beauty. Orchid lovers have dug them up by the thousands and transplanted them, usually without success, to their own gardens. Wild populations of these flowers are now becoming endangered, but unlike certain other imperiled species, these stand a fair chance of enduring. (See color plate 9.)

Lady's slippers deserve to survive, not only for the aesthetic instincts they arouse but also for the instruction they offer in such fields as floral evolution, ecology, and plant geography. As Linnaeus recognized, the bloom produced by a lady's slipper resembles a delicate foot or a finely crafted piece of footwear. The shoe is actually a modified petal known as the labellum. All orchid flowers have a labellum, but in the subfamily Cypripedioideae the labellum is an inflated pouch, white to gold or pink through purple in color, depending on the species. Framing the labellum are two more petals and three sepals, which lack the labellum's vivid colors. The dark, often ribbonlike petals and sepals have been compared to long shoelaces or thongs, and they usually bear most of the scent hairs (trichomes) on the flower. Some *Cypripedium* flowers have sweet, candylike aromas. The fragrance of the large flowers of *Cypripedium irapeanum* has been compared to expensive French perfume.

The scientist's goddess has often relinquished her name to the peasant's Virgin. Species of *Cypripedium* in Europe were consecrated by the French, Italians, and Germans, who call them Slipper of Our Lady. American colonists have been more playful, naming them whippoorwill's-shoe, camel's foot, hare's lip, old goose, ram's-head, Noah's ark, and moccasin flower. In Mexico one species is called *flor de pelicano,* or pelican flower, alluding to the pouch that would seem to hold more than its belly. Its floral anatomy may appear rather quaint, but it is functional. The sexual organs are atypical and are usually regarded as rather "primitive" for the orchid family.

Most orchid flowers have only one, fertile, pollen-making stamen, which is fused to the back of the pistil's neck, forming the pollen-releasing/pollen-receiving organ known as the column. The pollen-receiving surface, or stigma, of the column tends to be rather soft and gooey, forming a "quick sand" for incoming wads of pollen, called pollinia.

In contrast, the column of a *Cypripedium* flower has two fertile stamens. Directly beneath the stamens is a wide, domelike stigma, which is not a gooey swamp but is armed with tiny hooks. Directly above the stamens lies a shieldlike organ known as the staminode, which is divided into three lobes. The lobes of the staminode form a roof over the two stamens and the stigma and also turn the base of the labellum pouch into a lidded box.

Almost all *Cypripedium* species require some degree of cross-pollination to set seed, and our American slippers depend on bees. The "shoe," the column, and the column's lobed "lid" all play a vital role in manipulating the pollinator. Lured by color and perfume, a bee enters the labellum through a fissure on the surface and, once inside, browses on hairs lining the pouch. The bee is unable to leave the same way it entered, because the rim of the entrance fissure curls inward, so it escapes by crawling toward the base of the pouch where the labellum meets the column. There are two exit holes formed by the column and its staminode eaves. Each hole is "guarded" by a stamen. As the bee leaves the flower, it squeezes underneath one of the stamens, and sticky lumps of pollen are smeared across its back. When the bee visits a second *Cypripedium* flower, it brushes under the stigma and the stigma barbs, which rake the pollen lumps from the insect's pelt.

Shoe size reflects pollinator size. The flowers of the aptly named sparrow's-egg lady's slipper *(C. passerinum)* wear a labellum pouch

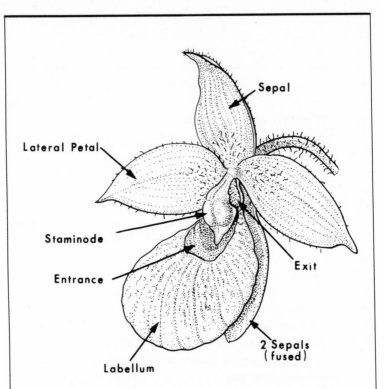

Lateral Petal

Sepal

Staminode

Exit

Entrance

Labellum

2 Sepals
(fused)

The flower of Cypripedium irapeanum *from Mexico shows the inflated, shoelike labellum petal. Bees enter the labellum through an entrance slit or fissure but can only escape through two exits on either side of the base of the staminode. The bee is forced to crawl under the stigma and the shedding anthers. Illustration by J. Meyers.*

that is barely more than ½ inch around. It relies on small solitary bees for pollination. The pink moccasin flower *(C. acaule)* offers a pouch that may be over three times the circumference of the sparrow's-egg. Most successful visits to a pink moccasin flower are made by the large, hairy bumblebee queens discussed in chapter 11.

Why would any bee fly into the trap of the *Cypripedium* flower more than once? Sometimes, when a small insect enters a very large labellum, it loses its way and dies, turning the maze into a crypt. Some bees never return to a *Cypripedium* flower after their first escape, yet some are willing to go again and again and are responsible for most cross-pollination. The bees do not eat the lumpy pollen smears of individual pollen grains loosely threaded together with sticky threads of elastoviscin, but they may eat the hairs lining the pouch. Some biologists suspect these offer the bee a "junk food" that is so stimulating, the insect returns again and again. I can't see a bee going through all this trouble for the orchid equivalent of potato chips. I've often wondered, then, if the hairs appeal more to a bee's sense of interior decoration than its palate. Bumblebee burrows tend to be cozy Hobbit holes. Bumblebee queens like their nests lined with soft, dry fibers of grass, fluffy animal hairs, and even furniture stuffing or the down from burst pillows. Some bumblebee species are known as carder bees because the queens are so often observed carrying insulation materials. They are so anxious to set up housekeeping sometimes that they will begin nest construction on the bare ground instead of an abandoned mouse hole. Someone should catch a few queens as they leave *Cypripedium* flowers to see if they are transporting the hairs from inside the labellum pouch, or perhaps someone should examine the insulation of a completed burrow to see if the orchid hairs are among the collected materials.

The subfamily, Cypripedioideae, contains three genera of slipper flowers that are confined primarily to tropical zones, *Selenipedium, Phragmipedium,* and *Paphiopedilum,* where, in fact, the vast majority of orchids grow. In contrast, there are between thirty and forty true species of *Cypripedium,* all of them found exclusively in the Northern Hemisphere, and most surviving well in colder climates. North America is blessed with about eleven species, two of which may be subdivided into several distinctive varieties.

So cold-hardy is *Cypripedium* that four species extend their ranges into southern Alaska. The spotted lady's slipper *(C. guttatum)* flowers in the spring woodlands of the Alaskan panhandle and the Yukon, but it is also distributed from Siberia to Moscow and southeast to Japan and northern China. We suspect that this species originated in the Old World, and the tiny, lightweight seeds made their way to our hemisphere by drifting on the wind and gradually colonizing the frozen island chains just south of the Bering Strait. Today there are still small, but healthy populations of spotted lady's slipper on the Aleutian and Kodiak islands as well as Honshu in Japan. The yellow lady's slipper *(C. calceolus)* also crowns the Arctic Circle but its range extends much farther south into the southern parts of the United States and Asia. It is also the most common slipper orchid growing wild in Europe.

Most species of *Cypripedium* have adapted to cold seasons by overwintering as dense, fibrous rhizomes. And while some species flower as early as April or as late as September, most plants avoid frost by blossoming from June through August.

Lady's slippers are also adapted to subtle variations in soil chemistry. Two North American species require the alkaline conditions of limestone swamps and prairies. The rest tolerate conditions ranging from acidic sphagnum bogs to the almost-neutral humuses of climax forests. Most populations are adapted exclusively to the microclimates and geology of their specific sites.

Lady's slippers have been described as "orchids that evolution improved, then forgot," suggesting that *Cypripedium* is a "senile" genus incapable of evolutionary change. This simply is not true. Lady's slippers show much the same genetic variation and mutability as other kinds of orchids in such traits as organ size, flower color, and stamen and labellum shape. In North America, hybridization is common between the yellow *(C. calceolus)* and small white lady's slipper orchid *(C. candidum)*. Varieties of yellow lady's slip-

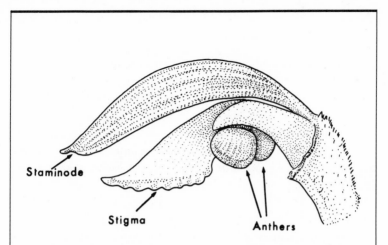

Staminode

Stigma

Anthers

The petals have been removed from this white lady's slipper
(Cypripedium candidum) *to show the organs that fuse*
together to form the column. The glossy staminode acts as
a loose "lid" over the scraping stigma and the pollen-
shedding anthers. Illustration by J. Meyers.

per cross with each other, as do the two varieties of spotted lady's slipper. This exchange of genetic material, as we know, may open the door to introgression (see chapter 10). The evidence suggests that *Cypripedium* is, in orchidologist Carlyle Luer's words, "a variable population of orchids undergoing active speciation."

The common belief that the lady's slipper orchids are not actively evolving is based on the fact that there are merely thirty species, which seems to be an unimpressive number compared with the hundreds of species found in tropical genera such as *Dendrobium, Bulbophyllum,* or *Oncidium.* These and other tropical orchids, though, are the beneficiaries of a favorable climate, which offers optimal conditions for growth and development and permits intensive speciation. *Cypripedium* species are remarkable precisely because they have been able to establish themselves so thoroughly in habitats unkind to most orchids.

The success of *Cypripedium* species in North America leads us to wonder why we don't see them more often. A century ago this question would never have been asked. The scientific papers of Asa Gray, the father of American botany; the journals of Henry David Thoreau; and the nature books of amateur naturalists all indicate that enormous populations were common at one time. In high summer these orchids dominated the protected bogs and isolated valleys of California, the Midwest, and New England. We now know that some of these colonies were not populations of individual plants at all. A solitary plant would form a huge, branching, rhizome that sent up many flowering stalks each year. What resembled a diverse population was really just a united clone. Since every flower in the clone was a genetic copy of its neighbor, rates of cross-pollination were low, and Asa Gray remained perplexed as to why such healthy "colonies" failed to set a single pod.

These days habitat loss is a major factor in the deterioration or loss of populations. This dilemma is as old as sodbusting on the prairies. An example is the small white lady's slipper. It is adapted to moist prairies and meadows, exactly the sorts of habitats that have been drained for agriculture, grazing, and new construction. The International Union for the Conservation of Nature has classified the small white lady's slipper as "vulnerable"; fourteen American states have listed it as rare or endangered. Almost all of the lady's-slipper species are in similar difficulty. This orchid's reproductive strategy works against it when its habitat is severely disrupted. A lady's slipper may take years to reach sexual maturity, putting most of its energy and food reserves into adapting to its respective habitat rather than into seed production. A colony of more than one hundred individual plants is likely to produce only about twelve seedpods a year.

With so many national parks and preserves, though, one would think that the decline would have self-corrected by now. Unfortunately, *Cypripedium* species suffer from an insidious problem shared by other plants that evoke aesthetic responses. During the International Botanical Congress of 1981, held in Sydney, I met a Dutch-born botanist who wanted to alert the scientific community to the gradual extinction of rare plants in Malaysia. The country is home to a number of forest preserves and other wild sites, but people were coming to these places, digging the plants up, and then trying to grow them in suburban gardens. Of course the rain-forest orchids, carnivorous plants, and bizarre parasites never survived, so the vandals returned to the same areas to take more. My Dutch friend referred to the process as loving a plant to death.

Unfortunately, enthusiasts have been loving lady's slippers to death since the early 1700s, when English nobility tried importing the rhizomes of American species to grow on their estates. During the last century there was also a demand for lady's slippers as pharmaceuticals and home remedies. Folklorist Vance Randolph recorded the ambivalence with which the people of the Ozarks treated their native slipper orchids. On the one hand, the plants were accursed. Some people believed they grew on spots associated with rural crimes and domestic tragedies such as rapes, miscarriages, and abortions. On the other hand, the local "yarb doctors" also believed that the rhizomes of *Cypripedium* contained a powerful aphrodisiac that could be released if they were boiled in milk. At various times parts of the *Cypripedium* plant were used to treat everything from pain in the joints to epilepsy. It was even recommended as a substitute for opium mild enough to give to children!

In the last century *Cypripedium* seemed to have been an American placebo with distinctly male chauvinist origins. While browsing at the library of the New York Botanical Garden, I discovered an American herbal written by a physician during the 1800s. He suggested that roots of pink or yellow lady's slippers be used as a tea or tincture to treat female hysteria and neuralgia. This sounds ridiculous, but such recommendations were believed, and in some old wildflower books *Cypripedium* species are commonly called female nervine. As late as the 1950s, powdered rhizomes of lady's slippers were still marketed as fluid extract of *Cypripedium*. I remember the back-to-nature crazes of the early 1970s, when most college towns had at least one natural food store complete with botanical remedies. Invariably there was one sad Mason jar filled with dried fragments of lady's slippers.

The harm caused by continued collection from the wild cannot be underestimated. By the turn of this century a number of nature writers were already beginning to complain that once-common species had been reduced to traces of their former abundance. It is amusing to note that these wildflower muckrakers often failed to practice what they preached. Grace Greylock Niles wrote in 1904, "There are laws protecting the deer in the Green Mountains and the brook trout in their spawning season, but as yet there is no legal or moral protection to shield the flowering and fruiting season of rare flowers, especially orchids, so scarce in northern New England." She railed against the drug trade in rhizomes and criticized people who foolishly tried to transplant wild orchids to their gardens. Of course, her own book, *Bog-Trotting for Orchids*, emphasized the pleasure of hunting wild plants, and it offers an early photograph of a veritable stack of freshly picked queen lady's slipper *(C. reginae)*, now one of our rarest species, ready to be pressed for "scientific study." She was a schoolteacher and often mentioned how her children would bring her lady's-slipper flowers as a variation on the apple-for-the-teacher gift.

We have plenty of laws today, but enforcement is difficult. It might shock a number of wildflower enthusiasts to learn that their hobby is most destructive, because it encourages commercial collecting. More than thirty years ago Donovan S. Correll, one of our finest orchidologists, reported that large clumps of lady's slippers could be purchased from dealers in native plants, but also noted that "colonies were rapidly being destroyed by this practice."

According to TRAFFIC (U.S.A.), the trade-monitoring office of the World Wildlife Fund (U.S.), entire populations of the pink lady's slipper *(C. acaule)* have been removed in a short time by large-scale commercial collecting. About four thousand of these are exported each year, and many more are sold within the United States. About three thousand plants of the ram's-head *(C. arietinum)* are exported annually. The California lady's slipper *(C. californicum),* may also be threatened by collecting. Both *C. acaule* and *C. californicum* are being studied for possible listing under the U.S. Endangered Species Act.

In its native habitat healthy *Cypripedium* rhizomes increase in number with each growing season over a period of years until they form a dense clonal mat. Taken away from the habitat to which it is specifically adapted, and faced with the shock of transplantation to a garden with a different soil chemistry, microbe flora, and climate zone, a plant can be expected to die. Stems will be injured by slugs and the rhizomes gnawed by mice as a source of winter starch. It's difficult to convince people that *Cypripedium* species will not thrive in their gardens. A rhizome usually sends up a flowering stalk the first year after transplantation, a few leaves during the second year, and nothing by the third. This has led many amateurs to believe that lady's slippers are biennial, like carrots!

Wildflower catalogs are often filled with deceptive language. They may describe their plants as "naturalized" or "introduced"

to cultivation. All this really means is that the plant was dug up
and transplanted to a nursery bed for a few months before being
sold to the optimistic gardener. At the time I write this chapter,
there are still no commercial sources for *Cypripedium* species
bred from seed or from true cultivated plants propagated artificially by dividing their rhizomes. Conservationists have been ex-
perimenting with *Cypripedium* seed for over forty years using
enriched soils and/or the agar-nutrient mediums I will describe
in chapter 15. Some species sprout just a few weeks after sowing
but then die after little more than half a year. *Cypripedium* species have a long "premie" stage in between the time the embryo
escapes its thin seed coat and the time it develops its first set of
organs. Even with optimal care, it takes at least three years for
potted seeds of the yellow lady's slipper to develop their first
functional leaves. Research in Canada has shown that the seedlings of other species can be raised only to a certain size, and
then they refuse to grow any further. It is also unlikely that
lady's slippers will become candidates for mass marketing from
tissue-culture suspensions (mericloning) in the near future, although this technique has taken some of the pressure off natural
populations of the Venus's-flytrap.

Only once have I seen lady's slippers flourish in a private
garden. The owners were fanatical growers of native plants. They
planted the dwarf variety of yellow lady's slipper (var. *parviflorum*)
in a bed reinforced with limestone blocks and then sent an artificial
stream cascading alongside. This was the only way the colony could
be sustained. How many gardeners are willing to invest this much
time and expense?

No wildflower gardener should live under the delusion that an
orchid in a pot or a garden bed somehow represents a saved species.
Such attempts at domestication are a matter of enthusiasm, hard
labor, and luck, it is true, but what sort of plans have these "conservationists" made for the time when they will be unable to care for
their specimens? Plants requiring continuous care are analogous to
auntie's heirlooms that no one wants, with the exception that a
piece of furniture can be consigned, for a while, to an attic until

a serious owner can be found. Lady's slippers in captivity cannot tolerate neglect, no matter how benign it is in principle. Botanical gardens and arboreta are often most reluctant to accept bequests since they don't have critical information on the orchids' points of origin. The orchids may harbor pests or disease, and, quite frankly, taking in such refugees only encourages the continuation of a vicious cycle.

Luckily the damage we have done is not irreversible. Protected remnants of old swamps and meadows may hold enduring populations that will increase, given time and privacy. I refuse to tell where, along a busy turnpike in upstate New York, I saw a clump of yellow lady's slippers *(C. calceolus),* but I will say that would-be pickers should note that most *Cypripedium* species bear hammer glands on their stems whose secretions may cause a nasty, poison-ivy type of rash.

The tiny, wind-dispersed seeds of the pink moccasin *(C. acaule)* will colonize maturing forests that have been replanted by conservationists. In Minnesota I noticed that pink moccasins did splendidly on land maintained by the U.S. Forest Service. A careful program of timbering trees of certain ages and height has allowed light to penetrate the forest canopy, encouraging the expansion of pink-moccasin rhizomes. In forty-year-old pine groves in North Carolina a few adult pink moccasins had time to seed and gave rise to several new generations of plants.

For those who still hunger for slipper orchids, the craving can be assuaged by the careful purchase of a *Paphiopedilum.* These inhabitants of Asia (from the Himalayas to the Philippines) are segregated from the genus, *Cypripedium,* by only a few reproductive and vegetative features, which are only important to taxonomists. *Paphiopedilum* species also suffer from overcollection, but careful shoppers will be rewarded by plants that have been grown exclusively from seed or represent decades of vegetative propagation in well-groomed greenhouses. In addition, some of the horticultural hybrids do quite well as houseplants.

It's time to rediscover the pleasures of admiring *Cypripedium* orchids in their native habitats and leaving them exactly where they stand. I'm a thirty-six-year-old botanist who has never dug up a single *Cypripedium* plant, and I see no reason to change my policy. If we'd exercise a little healthy respect and a little intelligence, we would ensure future generations a place to worship at the feet of Venus.

Chapter 14

San Salvador's Urban Orchids

Orchids are spoken of as the "weeds of the tropics."

—*Grace Greylock Niles*

THE CENTRAL AMERICAN country of El Salvador is the smallest mainland American republic, roughly the size of Massachusetts. Nevertheless, it possesses an abundant flora to delight and fascinate a systematic botanist. Orchid watching became one of my favorite weekend "sports" during the two years I lived in that country as a Peace Corps volunteer. One Sunday, for example, I found two orchid species in blossom. These sightings were not made under especially romantic circumstances. I was not hacking my way through rain-forest lianas or climbing a cow path along one of El Salvador's numerous volcanoes. I was walking down some of the busiest streets in the capital city of San Salvador.

The first plant, *Cattleya aurantiaca,* which produces small, star-shaped, red-orange blossoms, was clinging to a telephone pole. The second plant, *Oncidium cebollata,* was brightening the fork of a tree growing in a school playground. It bore a spray of flowers, each one the size of a child's fingernail. The sepals and petals of this species have a polished yellow finish marked with red-brown spots. Its common name is *chorizo con huevo* (sausage and egg).

As far as San Salvadorans are concerned, most of the ten orchid species that adorn their streets are as useless as the weedy plants North Americans associate with their own cities. We do not deliberately cultivate our urban dandelions, crabgrass, ragweed, or *Ailanthus* (the tree that still grows in Brooklyn). This vegetation simply flourishes under minimal conditions. Most city dwellers just ignore it.

Such plants are colonizers that benefit from the exposed soil of a disrupted or destroyed ecosystem, be it a fire-ruined forest or an abandoned lot. Their seeds germinate rapidly, and their growth pattern is typical of weeds: long, spreading stems; dense, durable root systems; and anatomical and physiological adaptations that permit them to withstand environmental pressures from abusive climates, insects, and grazing mammals. Most important of all, the energy that weeds glean from their habitat is almost completely directed toward their reproductive systems. The life spans of weeds tend to be short and nasty. They flower and die but the production of several generations each year ensures that there will be many offspring during the limited growing season. But there is a penalty to be paid for a life cycle in the fast lane: seed production must often be accomplished without cross-pollination, and this limits gene exchange.

We do not usually associate this life-style with orchids. In most cases, the growth genesis of an orchid plant must occur in the environment in which the original seed was deposited. Members of the family Orchidaceae grow primarily in climax ecosystems and old, stable habitats. The sunlight, water, and minerals orchids slowly convert into life functions indicate long maturation and complete adaptation to the subtle ties of their habitat. While a single orchid fruit may contain thousands, if not millions, of viable seeds, the plant may not flower for several years. Individual species rarely flower more than once a year. Compare that to the fecund dandelion. In the tropics, the blooming season of an orchid plant may last up to two months, allowing ample time for gene flow between individuals.

San Salvador is hardly Walden Pond. The estimated 1987 population was 484,000, and the city still suffers from all the ills that plague most capitals throughout Latin America. The large transient population is now aggravated by civil war. Overexpansion threatens the water table, and there is a full range of environmental and hygienic abuses, ranging from the discharge of raw sewage into local rivers to the absence of air pollution filters and noise control units on automobiles and buses. Yet in the face of all these obstacles, ten orchid species maintain standing populations.

At 2,238 feet above sea level, San Salvador is in the climatic range that local topographers and meteorologists call *el clima templado cálido,* the climate of the higher central valleys. Sixty-six to 78 inches of rain falls during the rainy-season months of June through October, and the average annual temperature is 75 degrees Fahrenheit. Where natural vegetation has been permitted to remain, plant geographers note that a low-altitude hot-zone flora and a high-altitude cool-zone flora coexist. The orchids of San Salvador represent both floras. Species such as *Oncidium cebollata* and *Catasetum integerrimum* begin their range in the dry forests of the Pacific coast that I discussed in chapter 1. *Epidendrum chacaoense, Brassavola cucullata, Epidendrum adenocarpon,* and *Laelia rubescens* become common in the savannas of the low valleys. All the other urban species are essentially cool-zone plants, but they do descend to the lower altitudes and mingle on equal terms with other orchids in the parks and side streets of San Salvador.

All urban orchids are epiphytes and exploit microhabitats that human beings have no intention of claiming for their own. They spend their lives clinging to the bark of trees, but unlike the strangler figs or mistletoes (see chapter 3), orchids never parasitize their hosts. In this respect, they are much more like the bromeliads of chapter 12. However, unlike the *gallitos,* epiphytic orchids have long, tangled root systems that encircle tree limbs and form a living net that traps rushing rainwater and holds dust and leaf debris, which can be used as a mineral supplement after it's broken down by epiphytic bacteria and fungi. The epidermis of these epiphytic roots is a spongy, often multilayered sheath known as the velamen. The velamen increases the surface area for each root, increasing its ability to absorb dripping water. The sheath may also provide a home for nitrogen-fixing blue-green algae.

Just as parks and tree-lined streets form islands for orchids in a sea of asphalt and concrete, much the same thing happens in the countryside. Roughly 80 to 95 percent of El Salvador's forests have been cleared for coffee (the country's most important export and greatest source of wealth). When this happens, the orchids of the herb layer, such as the many species of *Spiranthes, Habenaria,* and *Sobralia,* vanish for good. But coffee growers leave the tallest trees standing to shade coffee bushes, so that many epiphytic orchids are granted a reprieve. Local populations survive, if on a somewhat limited scale. In addition, at the onset of the rainy season, Salvadoran farmers cut boughs of local tree and shrub species, strip the branches, and use them as stakes for barbed wire. Many of the stakes take root and put out new branches. Within a few years these

Portrait of an urban dweller, Brassavola cucullata. *This species often colonizes the limbs of mango trees in the city of San Salvador. The orchid withstands the dry season by living off water stores in its succulent pointed leaves. The white flowers are tinted gold and green and release a sweet perfume as the sun sets. Illustration by J. Meyers.*

living fences and windbreaks become hosts to orchids and other epiphytic plants.

Only a few inches of rain fall during El Salvador's dry season, November to May, but since dryness is a cyclical aspect of the climate, the orchids evolved adaptations to deal with both the excess and absence of water long before the advent of urban blight. The leaves and shoots of all orchids are covered with a thick waxy cuticle that retards evaporation. The secondary leaf-bearing stems are often swollen and globelike, forming succulent and fibrous structures that botanists and orchid fanciers call pseudobulbs. As the name implies, pseudobulbs are not true bulbs. Orchid fanciers once confused the two different organs, much to the detriment of their plants, but that is a tale I am saving for later. Like most species of cactus, orchids growing in dry places draw water through their roots during the rainy season and store moisture in their pseudobulbs for the long months of deprivation. Of course, orchids growing on trees in areas of high moisture may also have pseudobulbs, because even if rainfall is plentiful, drainage on a tree limb is rapid.

These pseudobulbs are indicative of the generosity of the season. While rains fall from June through early October, they are plump, glossy, and they put out new leaves. By the end of April, though, the pseudobulbs of some species are wrinkled and sunken and have dull-colored or dried leaves.

Not only are El Salvador's orchids able to tolerate changes in climate, they are also able to thrive on a remarkable variety of cultivated and introduced tree species. I have seen the tiny flowered *Barkeria chinensis* and the handsome *Epidendrum ciliare* and *Brassavola cucullata* forming massed colonies on the Australian pencil pines *(Casuarina)* that line the road to the National Fairgrounds. The dense branches of the monkey puzzle tree *(Araucaria)* support the heavily pseudobulbed *Catasetum integerrimum.* Even the bottlebrush trees *(Callistemon)* on the grounds of the National University play host to naturally dispersed colonies of *Epidendrum adenocarpon.* Rubber trees *(Ficus elastica),* imported from South America, and coconut palms *(Cocos nucifera),* which are probably native to the tropical Orient, were among the few cultivated foreigners on which I rarely located a single orchid.

I did find orchids in parks, in playgrounds, along streets—in virtually any spot with trees. Along the banks of the Rio Lempa, though, the homeless had cleared the trees and scrubby vegetation from the ravines. There were no orchids to be found there—just mile after mile of flimsy *barrios*, with sad huts made of bits of cardboard, sheet metal, and a few adobe bricks.

While I never had the opportunity to study the situation thoroughly, there did not seem to be many barriers blocking the fertilization of city blossoms. If a microhabitat is large enough to support colonies of epiphytic orchids, there should be room enough for small populations of pollinating insects, which seem to be attracted by the floral fragrances. A friend of mine compared the odor of flowers of *Epidendrum adenocarpon* to a mixture of orange blossoms and burned rubber. The fragrance was strongest in the early morning, when the flowers were visited by small, solitary bees that carried off the sticky pollen wads (pollinia) on their backs. In contrast, *Brassavola cucullata* waited until the sun set to release a hauntingly sweet aroma that attracted hovering sphinx moths. Some moths were larger than hummingbirds, and I often watched them "warm up" against the hot concrete of city pavement while they vibrated their long wings. Both of these orchid species offer floral nectar to the insects as a caloric reward, but it is still estimated that more than half of the species in the family Orchidaceae produce nothing for their pollinators to eat. Consequently the relationship between some urban orchids and their insects involve much more subtle interactions than the usual "sips for sex" system.

The reproductive success of nectarless orchids can be quite startling in an urban environment. Highly evolved floral anatomy proves to be quite functional, even though the plants are so obviously divorced from their dry forests of origin. *Catasetum integerrimum*, for example, belongs to one of the few orchid genera with unisexual flowers. These plants are usually dioecious, which means that each plant produces either male flowers (the anthers are fertile, the ovaries small and useless) or female flowers (small infertile anthers, functioning ovaries), so self-pollination is impossible. The flowers of both male and female orchids lack nectar glands. Nevertheless, by the end of the flowering season most female plants have at least one fertilized pistil, which swells with seeds until it is almost the size of a lemon.

The male and female flowers of *C. integerrimum* exude a potent, chemically complex scent that is reminiscent of oil of cloves. During the morning hours this fragrance attracts large, furry bees in the genus *Eulaema*. The flowers provide nothing to eat, but the bees work hard gathering up the odor to store in "scent flasks" on their hind legs. Since the *Eulaema* bees visit male and female plants, some pollinia are transferred to receptive pistils. The transfer of pollinia to bee is very dramatic, and orchidologist-entomologist Robert Dressler has compared the floral mechanism to a slingshot! The pollinia in the male flower is connected to a little elastic stalk (stipe) that is stretched over a column knob so that the stalk remains under great tension. When a bee probes the flower, it touches one of two little "antennae" on the column, which releases the stipe-pollinia unit so quickly that it is shot at the bee's body and adheres by means of a quick-drying glue (viscidium).

Why do the bees tolerate this floral sadism? More interestingly, perhaps, why is *C. integerrimum* pollinated almost exclusively by male bees? Pollination biologists once thought that male *Eulaema* bees were "floral junkies," using orchids as a fix. The floral fragrance was regarded as a sort of "bee cocaine." Field observers remarked how the usually cautious insects seemed to lose discretion as they foraged for perfume. They were easy to catch in butterfly nets and seemed to fly more slowly after collecting their fill of perfume. Such interpretations, while colorful, do not appear to be true. More recent research suggests that some male bees require specific scents in floral scents to complete their life cycle, as they may use an orchid cologne to mark territories or to stimulate male swarming.

Other urban orchids do not require insects at all to be pollinated. These species have adopted weedy self-pollination. The flower buds of *Barkeria chinensis* often failed to open completely; I never saw a bee or fly visit them. The anther opened up inside the bud, and the pollinia dropped unceremoniously onto the waiting stigma. Huge colonies of this small orchid ornamented the outer twigs of trees planted around the American embassy. In fact, *Barkeria chinensis* produced so many tiny flowers that the combined weight of the burgeoning capsules would cause the delicate twigs to snap off on windy days. Orchid plants often littered the pavement and were trampled by hopefuls lining up for visas.

The floral biology of one orchid, *Cattleya aurantiaca*, appears to have altered under urban conditions. In neighboring Guatemala, this orchid's red-orange color and tubular flower are typical of bird-pollinated species, and orchidologists suspect that *C. aurantiaca* is pollinated chiefly by hummingbirds. *C. aurantiaca* also crosses naturally with the large, purple-flowered *C. skinneri*, in Guatemala, resulting in a medium-sized, delicately scented, pink-blossomed hybrid known as *C. X guatemalensis*. This hybrid is absent in San Salvador despite the fact that both parent species are found growing together on the same trees. The urban form of *C. aurantiaca* fails to open completely; the scoop-shaped petal (labellum) remains partially folded over the sexual organs. The labellum also lacks the characteristic red dots believed to advertise hidden nectar glands, yet every flower on the urban *C. aurantiaca* sets fruit. Presumably, although its bird-and-bee visitors are absent in the city, the orchid flourishes by self-pollination. Of course, a flower that remains closed is not able to receive the pollen of a neighboring species, and that means a dearth of pretty hybrids.

El Salvador's orchids, even those urban species that lack floral nectaries, have established a sophisticated relationship with the local ants. Below the flower bud of most species, on the flowering branch, but not on or in the flower, is an extrafloral nectary. Before the buds blossom, these nectaries secrete large, luminous drops of fluid, which are gathered by tree ants. The orchid's flowering branches remain free of sucking pests, such as aphids. Further experimental work will reveal whether the ants studiously keep the orchid branches clean on their own or whether the nectar is a bribe to keep the ants from tending their own aphid "cows" on orchid flowers.

Seeding presents little problem for urban orchids. Many epiphytes, including aroids such as philodendron and night-blooming cacti, produce fleshy fruits or seeds with oil glands or sweet, pulpy seed husks. These are eaten by birds, as well as by bats and other mammals, and are eventually defecated or regurgitated in fertile places, such as lichen-encrusted branches. This method of seed dispersal is alien to most orchids, including urban orchids, which have papery capsules or pods. The capsules may take a year to reach maturity, but they are filled with an incredible number of seeds that are practically microscopic—a seed consists of a tiny embryo and a transparent seed coat (testa) only one cell thick. In the city or in the wild, these orchids do not need animals to spread their seeds. Ripe capsules split open and the wind carries seeds away.

I found orchid plantlets citywide on both rough-barked and lichen–covered trees. Seeds fortunate enough to reach the latter havens germinate in the presence of certain strains of fungus, which nurture the epiphytic infant until it presumably produces its first leaves and roots.

Another factor contributing to the survival of orchids in the city is the apparent apathy of the general populace. Over three hundred species of orchids have been described in El Salvador. While I lived there, orchids were not generally worn to social functions, and the middle-class was just beginning to discover the possibilities of using such plants as part of the house decor. It's true that a number of the old Spanish herbals discussed potential medicinal uses for native vanilla orchids, but these wild plants remained uninvestigated and unexploited.

The Salvadoran attitude toward orchids is reflected linguistically. The word *orquídea* is a technical term used by plant fanciers and botanists in San Salvador. All tree-dwelling plants, including orchids, bromeliads, pepperomias, ferns, certain cacti, mistletoe, and strangler figs, are known by a single collective name, *parasita* (parasite). Few Salvadoran orchids have common names, and those that do frequently share them with members of the same genus. For instance, *chorizo con huevo* is applied not only to *Oncidium cebollata* but also to all yellow-and-brown-flowered members of the genus *Oncidium,* and there are fourteen *Oncidium* species distributed throughout the tiny country! *Epidendrum pentotis* and *E. fragrans,* flowers of the city and of forests and coffee plantations, are known collectively as *conchitas* (little shells). *Catasetum integerrimum* and *Clowesia russelianum* share the surprising name of *zapatellas suecas* (little Swedish shoes). The *shoe* part of the name must refer to the slipper-shaped labellum, which is also found in our North American *Cypripedium* species (see the preceding chapter), but I have found no good explanation as to why they suggest Swedish clogs in particular. There are only two or three more species or genera that enjoy a folk name. How unlike the British tradition, inherited by North America and most Commonwealth nations, in which a wild orchid may have common names that change with territory and ethnic diversity. Members of the Salvadoran Orchid Society tried to give common names to their favorite species. Unfortunately most of the time they were just Spanish translations of names invented by English-speaking horticulturists.

Of course, since the flowers of the urban species are rarely larger than an inch across and the plants are often hidden by the dense foliage of their host trees, they are comparatively safe from dilettante collectors and gardeners. *Oncidium cebollata* frequently grows on mango trees planted in the city to shade strollers. Green mangoes are popular sliced and served with salt and chili sauce, and the poor are adept at knocking the fruits down with sticks and stones. I never saw anyone apply the same technique to orchids that grew out of reach.

The case of one urban species indicates that, for orchids, popularity is risky. As late as twenty-five years ago, *Cattleya skinneri (flor de San Sebastian)* was a familiar sight in San Salvador, and books on the local flora note how common the flower was in the city. A comparatively recent burst of interest in what is now the national flower of Costa Rica has caused these extremely showy plants to vanish from most of their old haunts except the most inaccessible tops of ancient monkey-puzzle trees. The plants may still be found in San Salvador, but their distribution can hardly be called natural. Now they brighten the enclosed private courtyards of middle-class homes, where they are stuffed into window boxes or wired to branches and logs.

Many questions about urban orchids remain unanswered. Why, for instance, are some orchid species seen on the outskirts of San Salvador but never seem to penetrate the urban heart? Large fig trees mark important bus stops just outside the city limits. There I found *Epidendrum stamfordianum, Isochilus,* and dense, matlike colonies of *Hexadesmia* species. Were these plants merely remnant populations from a time when the area was a forest, or are there hidden environmental factors that keep them around the periphery and deny them a niche in the city proper?

It certainly does not take very long for orchids of the New World tropics to invade and colonize new sites. Three scientists recently examined the orchid flora of a grapefruit orchard in Belize (British Honduras). They counted nearly thirty-six hundred individual plants and a total of twenty-one species. The orchard was only thirty years old. The diversity of orchid species seems to correlate with the age of the grapefruit tree and the thickness of its branches. The older a tree becomes, the wider its branches; more orchid species can then find an appropriate "perch" within the canopy. Considering all the pressures on an urban tree, it's possible that most fail to survive long enough to develop the wider boughs that would accommodate more orchid species.

Will the terrestrial orchids be forever denied their old territory in the city? I found two species of these ground plants surviving in San Salvador. The year my Peace Corps contract expired, I came across a mature *Cyrtopodium punctatum* growing in a bed of snake plants *(Sansevieria)* at the National University. This orchid is the largest species in the country, bearing thick, rigid pseudobulbs more than 3 feet long. It's a tough plant with an extensive natural distribution from the tip of Florida (where it is known as the cowhorn or cigar orchid) down through Central America and into

Brazil, but it also spreads eastward from Cuba on through the Lesser Antilles. The individual I discovered may have been a survivor from the days when a portion of the university was really an autonomous institute for research in tropical biology. This orchid may have been the lone survivor of a former collection planted by German botanists.

On the other hand, there was no easy explanation for the tight little colony of green-flowered *Habenaria alata* I found the same year. They were growing in the center of a well-tended lawn. "Oh, those?" the owner said, "Aren't they cute? Just sprang up one day."

I suspect that orchids will continue to colonize San Salvador for as long as appropriate microhabitats remain and the interest of the residents remains tepid. One day I proudly dragged a girlfriend across town to see what I considered an especially fine stand of *Encyclia cordigeron (Flor de Incarnacíon)* growing on an African tulip tree off a side street. This was my favorite urban orchid, as the flowers smell like vanilla extract and the labellum petals range in color from pink to deep amethyst.

"How ugly," the young lady sneered, "so small and colorless. Not at all like the ones I got for proms."

Chapter 15

Orchidelirium

Plants are marvelously docile. When they die prematurely, it is not of "treatment" but of maltreatment, and with Orchids especially, as with women and chameleons, their life is the reflection of what is around them.

—*Edward Sprague Rand, Jr.*

DURING THE NINETEENTH century, western Europe and America contracted orchid fever. The fad to collect and grow orchids was so seductive, so exhilarating, and so addictive that it spread rapidly and had an extraordinarily long lifetime—rather like an irresistible epidemic. It also had an extraordinarily destructive effect on populations of native orchids. Unlike other crazes, however, orchid fever did not simply abate, but instead transformed itself into serious study of orchids. What began as a destructive form of conspicuous consumption has drawn the interest of some of the greatest horticulturists and scientists of the last century. It has also changed the face of modern plant science in most unexpected ways.

Orchid fever had its origins in the early 1800s, when live tropical orchids from South America reached Europe. Collectors grew them in "stovehouses," conservatories and orangeries made of densely painted glass penetrated by only a small amount of light, lacking moveable windows or ventilation and heated by coal fires in thick brick flues. Ignorant of the subtleties of equatorial forest habitats, collectors really believed that their glass pressure cookers, heated to lethally high temperatures and heavily polluted by fumes and soot, faithfully duplicated the jungle climate. Most species survived just long enough to flower, then perished dismally.

At first, the prevailing notion that orchids would flourish for only the most sophisticated and knowledgeable owners fed the vanity of plant fanciers, helping to recruit converts to the orchid hobby. Demand increased, more orchids were imported, and they continued to die, no matter how sophisticated their cultivators.

So, stovehouses were improved. External heat sources connected to hot-water pipes winding through the structures saved the plants from direct exposure to soot, carbon monoxide, and dangerously dry air. It was still too hot for most orchids, though. Misinterpretation of plant anatomy also proved to be fatal. In their attempts to cultivate tree-dwelling (epiphytic) orchids, growers mistook the green, swollen, secondary stems (pseudobulbs) for true underground bulbs, like those of an onion or hyacinth. Pseudobulbs were plunged into tubs of wet compost and rotting sawdust, where they became diseased and soon died. Alas, orchid growing seemed to be too expensive and rather futile.

Historians of horticulture record twenty years of decline in interest and orchid importation between 1832 and 1852. During this period, though, major advances prepared the way for a vigorous revival of the mania. Nurserymen learned that they could minimize their losses during long sea voyages by shipping sensitive plants in tightly glazed cases. These so-called Wardian cases spared the plants from salt spray and the destructive radiation of unfiltered sunlight. Wardian cases brought to eager buyers orchid genera of cool zones and higher altitudes, such as *Odontoglossum* and *Lycaste.*

As each new plant arrived, its image was admired in expensive publications of lavishly hand-colored plates. These volumes took a long time to produce, but they whetted the Victorian appetite for the exotic. My own favorite is a specialized folio called *The Orchidaceae of Mexico and Guatemala* (1837–41). Author James Bateman not only offered fine portraits with descriptions, he also delighted in presenting small woodcuts depicting life in the Spanish colonies of the New World. Subjects ranged from unusual fishing techniques to customs at religious festivals.

To see the progress that was being made in orchid culture, we must return to Chatsworth (the estate of the sixth duke of Devonshire, a lifelong orchidophile) as well as that genius of tropical horticulture, Joseph Paxton, whom I discussed in chapter 9. Paxton realized that wild plants would flourish outside their countries of origin only if their environments were duplicated with rigor. Remember his success with the giant water lilies. He decided that orchids from high altitudes and volcanic peaks wanted mist and cool nights. So he lowered the temperatures, opened the windows, and spread wet sand on the floors of the plant houses. He let tree dwellers remain attached to severed branches and draped damp, live moss over their exposed roots.

Oncidium ampliatum *bears the swollen, rounded, secondary stems typical of many tropical epiphytes. Small wonder that early growers in Europe treated these pseudobulbs like true bulbs or tubers and buried them in compost mixtures. Illustration by J. Meyers.*

Paxton set the trend, and the Industrial Revolution brought it to the growing middle classes. In 1845, coal and cast iron were cheap, and England had just repealed its glass tax. This was the end of the storehouse era and the beginning of the age of the greenhouse. Soon the most expensive items were the orchids themselves. The middle class lacked the property and funds to establish landscaped estates, but they could easily afford the price of one or more conservatories erected behind their townhouses. The new greenhouses, combined with Paxton's cultivation techniques, made the culture of tropical epiphytes almost easy.

In 1869 the golden age of orchid fever began. During the first half of the century, missionaries and sailors made money transporting orchid novelties from tropical ports. Now, these souvenirs no longer satisfied the great demand. British nurseries such as Veitch of Exeter & Chelsea and Protheroe & Morris of Cheapside forged a direct line to the tropics and supported a corps of professional orchid hunters. The lives of these explorer-collectors were often nasty and short. Some died in South America of fever or embarrassing intestinal complaints. Others, like Thomas Lobb, lost legs in the Philippines. Benedict Roezl, considered the "King of Orchid Hunters," ruled with an iron hook after losing a hand. His piratical appearance did not protect him from bandits, who robbed him seventeen times while he explored Mexico and Central America.

It's hard to feel much sympathy for these men, however. Orchid hunters prided themselves on how little they paid their guides and porters. It was believed that the most successful kept rivals off their scents by drawing phony maps, leaving false trails, and even threatening competitors with death. Roezl was a lovely piece of work. One story tells how he procured white-flowered forms of *Cattleya skinneri*. Catholic clergy encouraged rural Guatemalans to decorate the facades of their churches with these hardy plants. Roezl obtained his quota by bribing parishioners to pilfer their own houses of worship.

Orchid hunters frequently caused the regional extinctions of certain species. The correspondence sent to employers and the field notebooks kept by contemporary botanists reveal that when a profitable area was located, it was plundered. Whole trees were chopped down, and orchid-laced branches were packed into crates and baskets and transported out on mule. Some plants died of exposure on the docks while others perished en route, often victims of stowaway insects or rodents.

The orchids that survived were usually sold at auction and brought highly inflated prices. One American collector proudly admitted paying £2,211 14 s. for only twenty-eight plants! Exorbitant prices contributed to the popularity of orchids by giving the plants snob appeal. Prices fluctuated widely—a £700 price tag on a rare variety could collapse if orchid hunters discovered a new cache and flooded the market. Collectors often bid most heavily on unfamiliar plants that were not in flower at the time of purchase, because they knew they had a distinct chance of buying a species new to science, and many dreamed of having an orchid named after them. The English hungered, too, for rare forms of more common species. A few old catalogs from Victorian nurseries show that white-flowered forms of a species almost always commanded higher prices. No wonder Roezl was willing to brave the anger of priests and devout peasants.

Of course, when you read the literature of this period, you will be struck by how frequently enthusiasts insisted that orchids fulfilled the Victorian preoccupation with good taste. Some said orchid collecting was preferable to collecting paintings or porcelains. A flowering plant would be carried into the parlor to be admired by family and guests. Since the life span of the blossoms of some orchid species is measured in weeks, they made surprisingly durable knickknacks. American collector E. S. Rand summarized this elitist emphasis on aesthetics by insisting, "The Orchid flower is neither superficial or fugitive nor insincere." He dismissed, however, lilies and fuchsias as inferior flowers that fell to pieces. (See color plates 10, 11, and 12.)

This orchid idolatry formed the base on which serious scientific research developed. In 1826 John Lindley, whom many plant taxonomists regard as the father of modern orchidology, published a large monograph of the family Orchidaceae. He divided all known species into seven "tribes" according to the number of anthers a flower displayed and the way in which it arranged its pollen into discrete packets known as pollinia. The Lindley system has undergone many revisions, but the basis for his method of classification remains sound.

Many naturalists, at the time, found the sexual anatomy of orchids most perplexing. Why were there so few pollen-making stamens in an orchid flower? Why did the remaining fertile stamen unite along the back of the neck of the pistil and form a column? Some botanists felt this column was a sort of conduit for something like semen, which would leak into the pistil from the anther and fertilize the unripened seeds in the ovary. Orchid flowers had rather quaint shapes, but few people thought that the column or the labellum had much interaction with the environment.

To understand the special significance of orchid floral anatomy, we must turn to another great botanist, Charles Darwin. In 1862 Darwin published a small book with a long title, *The Various Contrivances by Which British and Foreign Orchids Are Fertilised by Insects.* Darwin used orchids to show how natural selection modifies the organs of flowers to give them unique, adaptive significance. He showed that while orchid flowers may resemble birds in flight, spiders, and dolls, their forms are not merely the caprice of a benevolent Creator. Certain forms tend to increase chances for cross-pollination. The labellum may bear a "beard" that functions as a landing pad for an insect. If the base of the labellum tapers off into a long spur, it is probably filled with nectar. Warts and "tumors" are really attractive structures to an insect that help position it under the arching column, which houses both pollinia and their gluey receiver. Orchids regularly trick bees, flies, and moths into playing Cupid for the immobile blossoms. An insect leaving an orchid flower may not carry a quiver filled with the arrows of love, but it often has one or more clusters of pollinia glued to its body.

Darwin's book inspired a new generation of botanists more eager to observe than to collect specimens. Enthusiasm for field-work spread to the colonies Europe maintained through some of the most orchid-rich regions of the world. Now, the orchids that came under scrutiny were not necessarily those with huge, gorgeous blossoms, cultured in suburban greenhouses.

I realize this is a most personal prejudice, but I think that the most significant contributions to orchid ecology and geography came from the Australians. R. D. FitzGerald, a gifted amateur, offered some of the earliest reports (1875–1895) of how orchids survived in the arid bushlands near Sydney. The reputation of orchids as fragile beauties didn't seem as convincing after Fitz- Gerald described how Parsons bands, *Eriochilus* species, rapidly colonized the faces of hillsides after the land had been stripped for the laying of railroad track.

In Europe the trend toward serious research of orchids began to change the nature of the hobby of orchid growing. Hobbyists soon learned that hybridization could be a most enjoyable parlor game of few rules and great rewards. Sticky globs of pollinia could be easily removed with the point of a pin, while orchid pistils seemed unable to recognize and reject pollen of different species. Consequently, widely dissimilar species were crossed to form ro- bust hybrids, which were often easier to maintain in captivity than either parent species. By 1865 John Dominy, of the Veitch firm, had produced the first successful hybrid cross between two species of *Calanthe*. He also discovered that crosses could be induced between different genera. The Royal Horticultural Society of Great Britain held its first orchid conference in 1885. Three hundred and fifty species went on display, testifying to both the rapacity of the orchid hunters and the success of culture techniques under glass. However, Dominy's experiments would doom an era by changing public taste.

The greatest obstacle to hybridization was raising the plants from seed. Victorians just could not understand how an orchid pod could be packed with thousands of microscopic seeds yet in the greenhouse the number of plantlets that survived was almost nil. Growers achieved some limited success by sowing the seeds among the roots of the parent stock. Watered carefully, a few sturdy plants might endure and be harvested in three or four years. But hybridization was often as expensive a proposition as collecting wild plants.

Today we know that wild orchid plantlets survive only after certain strains of fungi invade their cells. The fungus nourishes the developing plant, and it's not unusual for a seedling to remain in a leafless, rootless stage (protocorm) for years. Noel Bernard explained the true nature of this relationship in a paper published in 1899. Botanists had known for years that fungi infested orchid roots but had regarded such "inferior molds" as minor pests, like athlete's foot or ringworm.

Unfolding the subtleties of this symbiosis for commercial purposes took more than two decades. Biochemist Lewis Knudson theorized that orchid seeds could be grown under antiseptic conditions on a medium impregnated with the same micronutrients that the fungal threads procured for the seedlings in nature. Eliminate the fungus middleman, Knudson suggested, and orchids could be cultured on a grand scale.

By 1922 Knudson had synthesized his first orchid "baby formula" of sugar and inorganic compounds cooked and solidified in an agar jelly. Many of the recipes developed since then, some calling for a nutritious base of tomato juice or coconut milk, have proven to be equally effective. Seedlings sprout inside sterilized jars. Once they grow roots and functional leaves, they are deflasked and moved to real flowerpots.

The advent of controlled breeding meant that a flower could be sculpted and tinted to the buyer's preference. With these home-made hybrids gaining popularity by the end of the nineteenth century, the interest in wild species declined, as did their prices. Collecting wild orchids became a child's pastime. A case in point is Oakes Ames, whose lifelong interest in orchids grew, in part, from his childhood collection of wild slipper orchids. He went on to become Harvard's first professor of orchidology and stimulated interest in the economic potential of the natural products of tropical plants such as balsa wood and curare. A childhood spent among orchids may have unrecognized intellectual merits.

Not everyone was pleased by the waning of an era. One man

noted sourly that the new techniques would prove to be the "death of the collector and the birth of the cut flower merchant." This has long since proven to be both a truism and a prophecy. The Fluffy Ruffles, favorite orchids of zillions of high school proms, derive from dozens of wild species from Central and South America. The hybrid lines that meet in these corsages have also involved crossing and backcrossing at least three different genera; *Laelia, Cattleya,* and *Brassavola.*

The sterilization programs and nutrient media pioneered by enthusiastic people who just wanted to grow pretty orchids are the source from which sprang methods for cloning many different kinds of plant tissues. Of course, industrialization has taken old methods to new heights of sophistication, but the original link is still obvious. The *in vitro* culture of orchids from seed is really a forerunner of the new plant-science technology. Plant cells, or tissue slices of stem tips, suspended on an antiseptic gel grow into whole replicas when maintained under controlled temperatures, given appropriate light regimes and treated with extra hormones. Modifications of the same techniques may be used to grow orchids, soybeans, corn, and many new varieties required for research or horticulture.

Today, with modern technology encouraging mass production, orchids are available in abundance. The countries of Southeast Asia, primarily Thailand and Singapore, dominate a multimillion-dollar industry in orchids for the florist. Even hobbyists, armed with powdered orchid-culture formula, now have the option of transforming their kitchens into small-scale seeding and flasking operations. Some breeders and nurseries will sell mature seedlings, ready to be deflasked and potted up, so that the buyer can witness the whole range of variation in a single cross (providing he waits a few years for the young plants to flower). More impatient souls can walk into almost any large plant shop and buy a mature hybrid in flower, which costs less than a steak dinner with wine at a good restaurant.

I've often wondered what the serious hobbyists from the golden age of orchid fever would think of modern advances and tastes. On the one hand, they would probably be thrilled by increases in the sheer physical size of the blossoms and the introduction of new color ranges absent in nature. Modern hybrids have certainly not lost their Victorian aura of opulence and indulgence in these respects. Early collectors would be mesmerized by the luxury of growing almost every seed into a flowering plant, then selecting the finest representatives.

On the other hand, it's likely that the first generation of orchidophiles would be perplexed, then repelled by a modern grower's attempt to breed out some of the characteristics that make an orchid flower unique. Go to any club and you will find competitors attempting to follow national rules of good form designed to eliminate the "sway-backed," bilaterally symmetrical flowers of the ancestral stock in favor of flattened, nearly circular blossoms. One hybridizer of *Dendrobium* species proudly announced that he had bred out the "ugly raised callus" on the labellum petal. Of course, this hideous birthmark is one of the same floral features that fascinated an earlier generation willing to pore over each new orchid flower with a magnifying glass.

For better or worse, I live in an era in which most of the exoticism of an orchid display or exhibition is an artful illusion. I'm the freak at the orchid show who pulls out a hand lens or buries his nose in just a few flowers, grateful for natural ornamentation or a whiff of the wild scents. Breeding technology and superior cultivation have marvelous implications, but they also lead to a degree of monotony that keeps me hiking along forest paths out of the greenhouses.

Chapter 16

Charles Darwin and the Christmas Star

The *Angraecum sesquipedale,* of which the large six-rayed flowers, like stars formed of snow-white wax, have excited the admiration of travellers in Madagascar, must not be passed over. A green, whip-like nectary of astonishing length hangs down beneath the labellum.

—Charles Darwin

"YOU MEAN THAT thing's real?" My students treat their field trip to the Orchid Show at the Missouri Botanical Garden with a mixture of glazed tolerance and outright incredulity. Their greatest expressions of disbelief are always reserved for specimens of the Christmas Star orchid *(Angraecum sesquipedale)*. It is one of the few plants that produces flowers capable of banishing the catatonia of an undergraduate taking notes. Also known as the vegetable starfish, star of Bethlehem and rocket orchid, its floral face may be 8 inches across. It seems too large, rigid, and shiny for the blossom of a truly wild species. It's almost as if a sculptor had tried to carve the Archangel Gabriel and his trumpet from a piece of ivory and blended together the forms of both spirit and horn.

The Christmas Star is the best-known of the 125 species of comet orchid *(Angraecum)* of Madagascar. Most of these species pass their lives as epiphytes clinging to tree branches. The Christmas Star prefers the humid lowlands along the eastern coast of the island, where most of the forests have already fallen to agriculture and erosion and where it receives about 150 inches of rain annually. It is still fairly common in the wooded remnants that Hedvika Fraser recently described as "the juicy jungle of my dreams, every tree dripping with water and orchids." Despite its rather restricted and still shrinking distribution, the flowers of Christmas Stars have inspired noisy debates and startling observations for over 125 years.

Charles Darwin dissected fresh flowers of the Christmas Star provided by James Bateman, a famous orchidophile of the Victorian era. Darwin noted that the curved triangular mouth of each labellum petal led ultimately to an extremely long spur, which held a few drops of sweet nectar in its deepest reaches. Since the nectar of the Christmas Star lay hidden at the tip of this tube nearly a foot long, Darwin postulated that the flowers would have to be pollinated by an insect with a tongue at least 10 to 11 inches in length.

He nominated the tropical sphinx moths as the most likely candidates. After all, even the good old sphinx moths of England had tongues as long as their bodies, which they kept coiled under their heads when not foraging for nectar.

Both entomologists and orchid fanciers openly ridiculed the idea of a Madagascar moth carrying such a freakish drinking straw. Others felt that Darwin's extrapolations were evidence that his theories of natural selection and descent through modification were as overextended as the proboscis of his hypothetical insect. However, by 1901, forty-one years after Darwin's "folly in print," a sphinx moth with the appropriate tongue was indeed collected on Madagascar. Named *Xanthopan morgani* ssp. *praedicta,* its subspecies name fulfills Charles Darwin's "original prediction." I marvel at the insight displayed by the father of evolutionary biology, since he never went to Madagascar to watch sphinx moths in action. In fact, Darwin's understanding of how the flowers of Christmas Star would function was based on sound, deductive logic.

Darwin did not examine the flowers of *Angraecum* species until after he had passed at least one happy season examining the native orchids growing near his country estate. Some of the local species of fringed orchids *(Habenaria* and *Gymnadenia)* produced light-colored flowers that released a fragrance each evening. The flowers were tiny, but they displayed slender spurs with nectar at their tips. Darwin's son, George, visited these banks of orchids at night and collected moths from the flowers. A moth would carry off the pollinia wads on its tongue or eyes after probing for nectar in the spur. Comet orchid flowers, like the Christmas Star, bear much the same equipment as fringed orchids. Is it any wonder that Charles Darwin thought the system in comet orchids might parallel the situation observed in these smaller ones?

Biological explorations of Madagascar have continued right through to the present time. Some scientists think they can see a natural pattern developing. There are at least sixty species of sphinx moth native to Madagascar, some sporting tongues whose lengths can compete with that of the *praedicta* subspecies. At last count there were an estimated 250 species of orchids on the island offering stiff, whitish flowers with arched or dangling spurs. It would seem that sphinx moth pollination is a dominant theme in orchid evolution on Madagascar. Could these insects partly fill up a niche that is dominated by large bees or small flies on other continents? Some authorities think that the diversity of sphinx moths and spurred orchids on Madagascar represents a trend toward mutual selection in which flower and pollinator coevolve.

In the past the Malagasy government has issued stamps depicting a *praedicta* moth parking its tongue inside a Christmas Star blossom. Ironically no scientist has ever observed any interaction between the two species in nature. Conjectures regarding the extent of coevolution between the moths and orchids of Madagascar have had to be treated as just-so stories based on reasonable presumptions. This view has changed radically just over the past few years.

Four botanists from Sweden and Madagascar have conducted studies in the field documenting the interplay between sphinx moths and orchids with written observations and photographs. L. Anders Nilsson, Lars Jonsson, Lydia Rason, and Emile Randrianjohany first turned their attention to a smaller comet orchid known as *Angraecum arachnites* growing in forest patches along the Central Plateau. The ghostly white flowers open in November and emit a sweet odor at dusk. Each flower wears a 4- to 6-inch-long spur that may curve or loop as it develops. The drops of nectar in the spur tip contain about 13 percent dissolved sugar.

It appears that the only effective pollinator of this comet orchid is the sphinx moth *(Panogena lingens)*. The moth hovers directly in front of the blossom and inserts its 4½-inch-long tongue down into the spur. Orchid pollination is carried out when the moth recoils its tongue from the bottom of the spur. The tongue drags against the sexual organs of the column, which releases the two pollen balls hidden in the single anther. The paired balls are connected to a sticky plug (viscidium), which attaches the pollinia to the moth's tongue. When the moth takes nectar from another flower on a second plant, it extends its tongue again and deposits

The flower of Angraecum comorense *is not quite as wide as a silver dollar, but it still wears a long, hollow spur with nectar at its tip. Flower buds on a hanging branch must twist upward before they open, so the wide, white labellum points up and the spur down. Illustration by J. Meyers.*

the pollen balls on the second orchid's pistil tip, enabling cross-pollination to occur.

The team of scientists collected twenty-eight *Panogena lingens* that wore the pollinia of *Angraecum arachnites,* and they photographed one individual with eight pairs of pollen balls on his tongue. Although they collected dozens of sphinx moths species on the Central Plateau, only *Panogena lingens* carried the pollinia of this comet orchid.

Additional field studies on other tree-dwelling orchids of the Central Plateau turn up a very similar story. It doesn't matter whether it is just another comet orchid like *Angraecum compactum* or members of other spurred genera like *Neobathia grandidierana, Jumellea teretifolia,* or *Aerangis fuscata.* Each species offers pale, glossy flowers, since white petals probably stand out better in the leafy gloom of the night forest. Each flower bears a spur measuring from about 3 to 5½ inches in length, depending upon the species.

There may be more than two dozen species of sphinx moths flitting through the treetops. However, *Panogena lingens* appears to be the only resident insect that pollinates each of these five orchid species. You can find pollen balls from each one on the same moth. All of the remaining moth species either ignore the five species or steal nectar from the spurs without transferring any of the pollinia.

Panogena lingens, then, is "master" of an entire guild of white, spurred flowers on the Central Plateau of Madagascar. Without its nightly carousing, many orchids could not set seed. This degree of specialization becomes even more dramatic when the pollinator is examined in greater detail. A sphinx moth emerging from its cocoon is either a long-tongue or a short-tongue form, which stays that way for the rest of its life. Long-tongue forms uncoil a tongue that is up to 4½ inches long. A moth with a short tongue wears a proboscis only half that length. Less than 25 percent of all short-tongue forms trapped by the scientific team carried the pollen balls of the flowers they visited, compared with more than 80 percent of long-tongue moths. If a comet orchid is not visited by a long-tongue

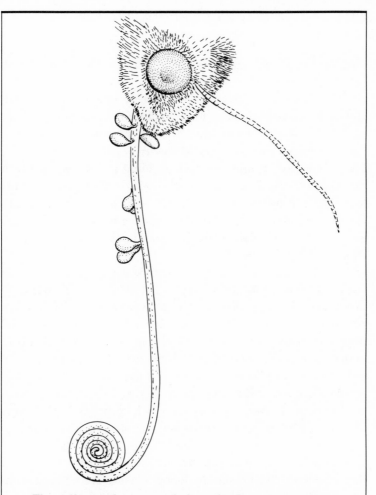

The pollinia of comet orchids and other angraecoids is deposited toward the base of the moth's tongue. When flying or resting, Panogena lingens *coils up its tongue like a party favor. Illustration by J. Meyers.*

form of *Panogena lingens,* its opportunities for cross-pollination probably undergo a serious decline.

This sort of "fine-tuning" is extremely rare in nature. Plant species are more likely to ensure reproduction by exploiting a wide range of potential pollinators. After all, even the most efficient insects undergo natural lapses of fidelity to their favorite flowers. It's still too early to draw conclusions, but it would seem that both the limited distribution of an orchid species in Madagascar and the long length of its spur would narrow the number of moth species capable of pollinating its flowers. A species of comet orchid may evolve like Alice's Red Queen, running in place just to keep up with one form or race of moth.

So, if it's any consolation, the orchids make *Panogena lingens* work very hard for its nectar supply. The orchid's spur tends to be just a little bit longer than the longest moth tongue. The insect is forced to push its head deep into the floral tunnel so that the sexual organs of the column come into direct contact with the thicker base of the tongue or the moth's broad, fuzzy face. The moth must push and shove to reach its nightcap, and this exerts enough direct force on the column to release the sticky plug and its paired pollinia.

Of course, Darwin hypothesized way back in 1862 that flowers like the Christmas Star manipulated their moths in this fashion! He conducted a series of simple experiments on the Christmas Star by passing thin straws or bristles over their columns and noting that thin structures, like the thin portion of a moth tongue, were insufficient to trigger the orchid organ, known as the rostellum, which harbors the sticky plug. Dr. Nilsson has recently conducted his own experiments using the fringed orchids of Europe, shortening their spurs by artificial means, then exposing them to thirsty sphinx moths. His results confirm Darwin's interpretation of the evolution of the "tug-of-war" relationship between moth and Christmas Star. In fact, these floral mechanics probably apply to most orchid species pollinated by moths and butterflies.

Don't you just hate the sort of people who seem to be right all the time? Every generation of naturalists rediscovers the wisdom inherent in Darwin's little book, paying its author the ultimate compliment by illuminating one of the more controversial passages. Frankly I don't understand how Stephen Jay Gould, of all people, could refer to this book as Darwin's "most obscure work." Even if you choose to disregard its long-term effect on botany and ecology, it's still responsible for planting unique monsters in modern fiction. We will save the details of that shocking disclosure for the next chapter.

Chapter 17

The
Fiendish Orchid

There were in it metaphors as monstrous as orchids, and as subtle in colour.

—*Oscar Wilde*

ORCHID FANCIERS ARE such innocents. They still believe that their hobby is completely safe. They ignore the peril of the orchid monster, a creature now more than ninety years old that has been described by some of the best-known writers of stories of the fantastic and the macabre. This loathsome and menacing orchid, assert these authors, may be lurking in some half-forgotten corner of the greenhouse. Not only is it a bloodsucker, but a recent novel suggests it is waiting to subject your loved ones to a fate far, far worse than death!

Villainous plants are nothing new, but most tales of folklore and literature implicate trees. We can follow a tradition as ancient as the vengeful dryads of Greek myth to the cunning of Old Man Willow in *The Lord of the Rings*. The orchid monster, however, is almost entirely a product of the Victorian era, when members of the family Orchidaceae influenced fashion and scientific theory.

It's quite plain, though, that proper Victorians established a love/hate relationship with their plants. The lurid colors, fleshy petals, and extravagantly formed flowers of orchids were treated as fascinating yet disquieting objects. For example, Oscar Wilde's novel *The Picture of Dorian Gray* introduces wicked Sir Henry Wotton, who proclaims that some orchids are "as effective as the seven deadly sins." In contrast, the genteel "language of flowers" treated the familiar frog orchid *(Ophrys)* of British bogs as a symbol of Disgust. Taxonomists relished the dark, clawlike blossoms of tropical orchids, giving some generic names such as *Mormodes* (Greek for "frightful object," or "phantom") and *Mormolyca* ("hobgoblin"). Applying spooky names continues to the present day. South America is now the home of the recently revised genus *Dracula,* and there is a species named *Dracula vampira.*

Certain extremists have enjoyed blaming Charles Darwin for most of the modern ills of Western civilization. In fact, if we take one last look at Darwin's little orchid book with the long title, we could convict him of providing the literary compost that cultivated the orchid monster. In Darwin's day his book became popular reading and went through several editions. We've seen how his speculations on the functions of floral anatomy could both irritate and enthrall different levels of society.

It is not surprising that Darwin's book came to the attention of the scientifically trained Herbert George Wells and sparked his fertile imagination. In 1895 Wells published a group of short stories, known as science romances, which included an odd little piece called "The Flowering of the Strange Orchid." In this story the protagonist, Mr. Winter-Wedderburn, buys the root of a mysterious orchid whose former owner perished under romantically gruesome circumstances. His corpse was found in an equatorial mangrove swamp, drained of blood, "with one of these very orchids crushed up under his body."

Winter-Wedderburn installs the root in his steamy greenhouse, and much to the distress of his housekeeper-cousin, the root grows into a robust plant producing a web of gently waving aerial rootlets that look, as the cousin says, "like fingers trying to get at you." When spikes of flower buds appear, the proud and excited horticulturist remembers his Darwin and conjectures as to the function of the unopened blossoms: "Some of the cypripediums, for instance; there are no insects known that can possibly fertilize them and some of them have never been found with seed."

When Winter-Wedderburn's beautiful flowers open, they release an overpoweringly sweet stench, which renders him unconscious. The deadly rootlets settle on him to drink his blood. He is saved only by the quick thinking of his cousin, who tears the roots from his face and hurls a brick through the glass of the greenhouse. Icy winter winds clear the room of narcotic fragrance and finally kill the tropical beast.

Wells's short story was the first of many stories about orchid menaces. These creatures, growing from the pens of principally British authors of science fiction and horror, have borrowed heavily from Wells's first descriptions. The situation is comparable to the way in which medieval herbalists plagiarized each other for details concerning the lore of the mandrake until information on the real plant became unrecognizable.

John Collier, whom film buffs may recognize as the scriptwriter for *The African Queen,* is the respected author of a short story, "Green Thoughts," that owes much to Wells and to some lines of poetry by Andrew Marvell:

> Annihilating all that's made
> To a green thought in a green shade.

Collier's story describes an orchid that prefers meat to blood and devours flies, a pussycat, an eccentric female cousin, and eventually, with the aid of tentaclelike tendrils, its owner, Mr. Mannering. The heads of the hapless victims emerge as flowers on the vine's stems, with whiskers, eyebrows, and beards transformed into fringed petals. Collier is deliberately vague about how the orchid ingests its victims, but he describes the transformation of human into blossom as a slow, dreamlike process.

The orchidaceous Mr. Mannering can still think and feel. He experiences pollination when a bee invades his greenhouse and crawls into what was once his mouth, but pleasure changes to mortification when the bee leaves him to fly to the flower that was once his prudish cousin.

Mr. Mannering's idyll is literally nipped in the bud when his wicked nephew discovers his uncle's new incarnation and finishes him off with a pair of scissors. Collier concludes, "In the vegetable world, only the mandrake could voice its agony—till now."

In "Prima Belladona," written in the 1960s, British author J. G. Ballard (now well known for his autobiographical *Empire of the Sun*) postulated a future world where people are so in love with the bizarre that they breed flowers to sing. Those without taste or cash can only afford daisies that warble pop songs, while the more affluent shop for floral divas. The Khan-Arachnid orchid, derived from an imaginary species from the Guinea jungles and living in a tank of fluoraldehyde and chlorine gas, lures pollinating spiders into its open blossoms by means of a hypnotically vibrating calyx. The hybrid of this orchid makes its own breeder permanently deaf.

The Khan-Arachnid hybrid in this story is the overbearingly proud queen of a florist shop selling operatic plants. The orchid, which tunes every plant in the shop, naturally cannot bear the presence of a beautiful human soprano, whose talents rival its own. Swelling to three times its ordinary size, the orchid tries to sing the girl to death, fails, and dies in a fit of pique the following day.

Ballard's story suggests that Frankenstein's monster might be assembled just as easily in a potting shed as in a dark castle laboratory. British writer John Wyndham took this concept one step further, implying, in his early 1950s novel, *The Day of the Triffids,* that harmless plant species could become the parents of foul, new creatures. This is perhaps the basest role in which orchids have been cast. At least the orchid monsters invented by Ballard and Collier did not reproduce.

Wyndham's novel is a tour de force in the way it incorporates a discredited theory of acquired characteristics known as Lysenkoism. Soviet biologist Trofim Denisovich Lysenko believed that altering the environment of an organism would permanently change the offspring's characteristics. In Wyndham's story, Soviet science develops the triffid as a means of coping with the ever-increasing demand for edible oils. Western corporations covet these plants and hire a South American industrial mercenary to smuggle out some seeds.

Señor Umberto is cryptic as to the origin of the triffid. "I have seen a picture, señor. I do not say there is no sunflower there at all. I do not say there is no turnip there, I do not say there is no nettle or even no orchid there. But I do say that even if they were all fathers to it they would none of them know their child. I do not think it would please them greatly either."

While Señor Umberto does not identify the plant's origin, Wyndham does tell us that triffids produce thousands of nearly microscopic seeds, so tiny that collectively they seem to resemble a white vapor. Anyone who has ever watched the pod of an orchid split open and the trickle of "seed smoke" disperse on the air current will recognize the unmistakable orchid paternity of the triffid. Triffids can also amble about on three stumps, and they possess a wicked, fatal sting. Although they supplement their diet with insects caught in their sticky funnel stems, they start to prey upon humans after mankind is blinded by an unexplained cosmic explosion.

Given the carnivorous viciousness of the orchid of literature, it could only have been a matter of time before someone tried to harness the plant for homicidal purposes. "The Reluctant Orchid" is possibly one of the finest and funniest of the shorter works of Arthur C. Clarke (author of *Childhood's End* and scriptwriter of *2001: A Space Odyssey*). Mild-mannered and puny orchid-doting Hercules Keating feels terribly dominated by his crass, overbearing Aunt Henrietta, a massive six-footer who raises German shepherds, drives a Jaguar, and hunts elephants in Africa for fun.

In the fashion to which we've grown accustomed, Hercules acquires an unidentified, shapeless root, which matures into a giant meat-eating orchid that breaks broom handles with its tendrils. Hercules believes he has acquired the perfect weapon for instigating an "accidental" death. He even goes so far as to starve the plant for two weeks to whet its appetite. The results are unexpected. Confronted by this amazon of a woman, the orchid suffers a complete nervous breakdown and is converted into another of Aunt Henrietta's "little pets." Hercules never recovers from the shock of his aborted crime and succumbs to a permanent stupor.

What is the underlying symbolic power of these evil orchids? They are a seductive, alien beauty brought into the homes of the jaded middle class. Their exotic charm is only illusory, however, since the orchids disrupt the smugness of comfortable citizens and society. The thread of mockery that runs through these stories is obvious. Orchid connoisseurs are portrayed as idle, flabby bachelors, well past their prime and—in devoting their time and energy to odd, finicky plants—not quite normal. Therefore they get exactly what they deserve at the hands (or tendrils) of their hostile charges.

H. G. Wells voiced a distinct impatience for anyone who would give himself, heart and mind, to a "tropical vegetable." In *War of the Worlds,* while Londoners flee the onslaught of the Martian war machines in terror, the hero-narrator comes across a shriveled old fellow with a huge box containing more than a score of potted orchids. Since the old man stubbornly refuses to leave his box behind, the narrator must grip him by the arm:

"Do you know what's over there?" I said, pointing at the pinetops that hid the Martians.

"Eh?" said he, turning. "I was explainin' these is vally-ble."

"Death!" I shouted. "Death is coming! Death!"

But, even as the narrator leaves, the old orchid fancier is still standing by his box and staring vaguely over the trees.

Wells and his cohorts emphasized the theme of decadent orchid culture, and for humanity to end up as lunch for an orchid was a perfect metaphor for the dying greatness of fin-de-siècle England. It could be argued that in our own time the appeal of these lush but predatory exotics springs from our own guilt over our affluence, which we may feel deserves to be punished. The predatory orchid is less common than the faddish gremlins and demonic children, but that doesn't mean it is any less compelling.

By 1970, at the height of the Vietnam conflict, flowers were symbols for everything at odds with a highly technological, hawkish society. Orchids became one of the more elegant representations of "flower power." John Boyd's 1970 novel, *The Pollinators of Eden,* offers a planet full of giant orchids just waiting for the right human to come along. These orchids have the usual arsenal of tendrils and absolutely devastating perfume, but the tendrils prefer to caress, and the floral fragrance is a subtle aphrodisiac. Botanist Fred Janet Caron, persuaded by her lover to commune with nature, falls in love instead with the new planet's flora. Her blood becomes mingled with the ichor of the orchids and, following an explicitly erotic interlude with one of these lusty plants, she gives birth to a seed that smells of vanilla but has the hair of her human lover.

Boyd's human pollinator may be a sign of a new relationship between man and orchid. More likely, it indicates a temporary truce in a protracted and bloody confrontation. In this current era of unresolved conflicts, modern orchid fanciers should follow a few simple precautions in order to enjoy their orchids in safety.

First, never accept a tropical tuber or hybrid plant whose collector or developer met with an unusual accident in foreign climes. Such roots and plants are identifiable because they are always redolent of carrion or decay. In the words of Winter-Wedderburn's cousin, they look "like a spider shamming dead."

Second, beware of any shoot whose aerial rootlets or tendrils sway in a breezeless greenhouse.

Third, trust the instincts of wary housekeepers and other dreary but tough-minded women. Above all, never forget that a flower is always an instrument of seduction. Wear a gas mask and earplugs in the greenhouse. If the plant should so much as belch, consign it to the compost heap. It may be exquisite, but it's not worth your life.

Chapter 18

Orchids in the Dark

One of the many inconveniences of real life is that it seldom gives you a complete story.

—*W. Somerset Maugham*

IN JANUARY 1979 the state of Western Australia began celebrating the 150th anniversary of its statehood. The event required a catchy slogan, of course, but the chosen phrase, "The State of Excitement," became a joke in the other Australian states. What did a bunch of "sand gropers," a less than affectionate term for the citizens of that state, have to offer that was so exciting?

The fall of Skylab, however, in an obscure corner of the Western Australian outback converted many of the scoffers. The spectacular rediscovery of an endangered species of orchid engaged the curiosity of both botanists and naturalists from all over the world.

The story behind the two Australian species of underground orchids predates Skylab by more than sixty years. During the Australian winter of June 1928, farmer John Trott cleared a section of his property near the small town of Corrigin by burning the dense thickets of mallee eucalypts and tea tree *(Leptospermum)*. While plowing the scorched earth, Mr. Trott's cultivator went through some dead, buried stumps of broom honeymyrtle *(Melaleuca uncinata)*. This unearthed peculiar, pale plants bearing single, daisylike flowers that seemed to have bloomed entirely underground. Trott dutifully turned his "daisies" over to C. A. Gardner, a botanist from the Western Australia government.

Gardner recognized that the plants were actually a bizarre orchid new to science. He enthusiastically made many drawings of the plant parts and sent his two specimens to R. S. Rogers, one of Australia's most eminent orchidologists. In this way the first orchid producing an underground flower received a technical description and was introduced to the scientific community. The genus was named *Rhizanthella—rhiz* meaning "rootlike structure," *anthella* meaning "little flower"—and the species name, *gardneri*, commemorated the helpful Mr. Gardner.

Until less than a decade ago almost everything we knew about *Rhizanthella gardneri* came from Rogers's original publication. He noted that the "daisy flower" is actually a collection of tiny, succulent red-and-purple blossoms assembled in a tight head known as a capitulum and enclosed within a cup of wide, pink, petallike bracts. The capitulum is mounted on the tip of a slender, scaly stalk called a peduncle. The peduncle, in turn, is attached to a thick, brittle underground stem (rhizome). The mature rhizome may be located as much as a foot below the soil surface and is not a very elaborate organ by any standards. It wears a few, thin "leaves," which have been reduced to nonfunctional scales, and a coat of white hairs. The entire rhizome is totally devoid of chlorophyll, the green food-making pigment, and there are no true roots, although the rhizome of a mature *Rhizanthella gardneri* may produce a few smaller "branches." Rogers thought these branchlets were flower stalks that had failed to mature properly and had aborted before a flower head developed. Today we know that these branches are young stem buds produced by the mature rhizome, known as "daughter tubers." After the rhizome flowers, it dries up. The next season's growth is carried on by the swelling and expansion of one of the daughter tubers. The orchid appears to creep through the soil, producing a new segment each year while the older portions shrivel.

Rhizanthella gardneri was the first but not the last subterranean orchid Australia would give to the world. Three years after the discovery of the Western Australian species, a second underground orchid appeared in the eastern state of New South Wales, 3,000 miles away from Corrigin.

The plant was found on the western slope of the mined Alum Mountain, growing in apparent association with the rhizomes of the hyacinth orchid *(Dipodium punctatum).* Five specimens of this new species eventually found their way to the Reverend H.M.R. Rupp, one of the greatest Australian authorities on the classification of ground orchids. Rupp realized he had a species similar to *Rhizanthella* in growth form and development, but the anatomy of the flowers differed. Rupp named the waxy white plant *Cryptanthemis* (*crypt* meaning "hidden," *anthemis* meaning "flowering head") *slateri* (for its discoverer E. Slater).

Almost every piece of original information on *Cryptanthemis* derives from the three papers Rupp published on the plants he studied. That is because *Cryptanthemis slateri* has been found only three times in this century: twice during the 1930s and once in 1974. Rupp's contributions are fascinating, not least because they reveal the fallibility of a dedicated naturalist: he was captivated by an orchid that repeatedly deceived him.

Rupp apologized for the poor condition of his specimens, yet delighted in documenting changes in color and structure as the flowers collapsed before his eyes. When he found that some of the characteristics he described were incorrect because his flowers were too old, he painstakingly redescribed them, meticulously redrew them, and published another paper acknowledging his past errors.

Rupp described white flowers that gradually acquired red splotches of color. Like *Rhizanthella, Cryptanthemis* is totally lacking in functional leaves and roots. *Cryptanthemis,* however, is branched like a candelabra—each of the three or four stems ends in a mature flowering head. The flower heads Rupp described opened a full ¾ inch beneath the ground. Rupp believed that after the little ovaries were fertilized, the flowering stalk pushed the ripening fruits upward until they broke through the soil's crust. He also believed that the fruits split open and the seeds were carried off by a gentle breeze. We will see, though, that *Cryptanthemis* may have fooled him again by masking its true method of seed dispersal.

Underground orchids survive in the total absence of sunlight. Their success has to do with the manner in which the orchids harness the food-gathering properties of the fungal threads with which they associate. The initial stages of germination and inoculation of underground orchid seeds are similar to those in regular orchids: fungal threads invade the embryo and form coils in the orchid's cells.

Once most orchids produce their first green leaves, the relationship between the seedling and the fungus alters. The young plant is no longer dependent on the fungus for all its nutritional needs and water supply. In other orchid species the symbiosis of the adult orchid and fungus disintegrates whenever the orchid enters a resting phase. These orchids must be "reinfected" every growing sea-

son while they develop new stems and roots. However, there is a
trend toward extending the extravagant dependency of the plant on
the fungus throughout the life cycle. Such orchids become complete
"mycotrophs," depending exclusively on the organic nutrients,
mineral salts, and water their fungal servants glean from organic
matter in the environment. The fungal threads that branch out from
the body of the orchid may be either saprophytes breaking down
dead plant material or parasites that attack the roots of healthy
living plants to steal food for their orchid masters. We have our
share of mycotrophic orchids here in North America. There are, for
example, a dozen native species of chicken toes and coral root
orchids *(Corallorhiza),* which produce slender, aerial sprays of
flowers from warty, leafless rhizomes. Australia has many orchid
species that are obligate mycotrophs, but these, unlike the underground orchids, always bloom aboveground.

It has been only within the last three to four years that the
subtle relationship between *Rhizanthella* and its fungus has been
well defined. The white hairs massed on the underground stem of
the orchid are natural conduits through which fungal threads enter
the body of the plant. The threads penetrate the thin cell walls of
these hairs and then bury themselves up to ten cells deep in the
storage tissue of the stem, eventually penetrating a storage cell and
forming a dense coil in the cell sap. Meanwhile, the fungal threads
still free in the soil continue to grow and branch until they reach
the roots of the shrubs of broom honeymyrtle and a few eucalyptus
trees. Fungal threads may be so dense that they clothe both the
orchid and the shrub roots in a moist, violet web.

When botanists first examined the orchid back in 1928, they
tried to isolate the fungus and concluded that it was a species of
Aspergillus. This was quite a surprise, as *Aspergillus* is a rather
notorious genus within the sack fungi *(Ascomycota)* and is known
for the often fatal diseases it causes in seedlings, bees, birds, bats,
and people. This identification, however, seems to have been the
result of an error made during the incubation technique, as experiments conducted in 1985 proved conclusively that the underground orchid really required a much more conventional strain of

Rhizoctonia. This strain of mushroomlike threads activates and supports many other species of Australian orchids.

The *Rhizoctonia* that lives inside the storage cells of an underground orchid does not invade the storage cells of the shrub roots it encounters. Instead, the fungus forms a sheath around the epidermal cells of the tree or shrub roots known as a Hartig net. It appears that the fungus steals mineral salts from the roots of the shrub and then transfers them to its coils inside the orchid. Once the fungal coils are plump with nutrients, the orchid cells kill them and extract the digestible treasure. The dead hyphae remain in their orchid cell coffins as black clumps. The orchid's ruthless betrayal does not discourage the fungus, however: hyphae reinfect the same cells over and over again, only to be similarly dispatched. The *Rhizanthella-Rhizoctonia* relationship puzzles mycologist Harry Swart of the University of Melbourne, who says, "It's difficult to understand just what the fungus derives from that relationship. The fungus invades the orchid, just as it would try to invade other root systems, and gets trapped by the orchid."

In spite of its murderous ways, however, *Rhizanthella* cannot survive without its *Rhizoctonia.* It takes from two to nine months for a *Rhizanthella* protocorm to escape its seed coat. Although a *Rhizanthella* protocorm may shed its coat without the aid of fungal threads, it soon dies unless proper penetration by *Rhizoctonia* takes place. Once the three-way relationship between orchid-fungus-shrub has been established, *Rhizanthella* grows very fast. A protocorm can become a mature flowering plant in just fifteen months.

The fungal relationships of Rupp's underground orchid *(Cryptanthemis),* have been harder to explain, since specimens are unavailable for study. One feature we have some knowledge of is its "association" with the hyacinth orchid. This is a rather mysterious association because the hyacinth orchid is also a leafless mycotroph. It shows a more conventional pattern of growth, however. The hyacinth orchid is a vandoid like the Christmas stars and the *Lycaste* orchids and produces a thick, tall stalk covered with large, purple-spotted flowers. It has been proposed that the fungal threads enslaved by *Cryptanthemis* are parasitic on the fungal threads enslaved by the hyacinth orchid. Rupp himself was skeptical about such a four-way relationship and suggested that *Cryptanthemis* and the hyacinth orchid share the same fungal strain by coincidence. In Australia *many* species of orchid seeds will germinate when exposed to the same fungal strain. However, the parasitism theory of *Cryptanthemis* still requires scientific refutation.

The two species of underground orchids have had a fairly successful career escaping botanists. The soil in which Rupp's *Cryptanthemis* specimens grew was not the original soil of the mountain slope, since the area was a waste dump from an old mine. Seeds probably either arrived in this debris or from mud slides following heavy rains. Further mining destroyed this orchid's location. Its rediscovery in 1974 occurred in the Blue Mountains near Sydney when a property owner excavated soil near his house to put in a garage. It has not been seen since.

As for *Rhizanthella gardneri,* it has proved to be just as elusive and rare. Until 1979 it had only been collected seven times. While this species appeared to have a fairly wide distribution in Western Australia, it seemed to occur primarily within the borders of the wheat-growing belt; more and more of its natural habitat came under cultivation every year. The last collection was in 1959, and then rumors began to circulate that this species must have succumbed to intensive farming and was now extinct. In hopes of adding to the pitifully small collection of preserved material, Alex George, curator of the Royal Botanic Garden of Perth, Western Australia, announced that the next underground orchid discovered would carry a one-hundred-dollar bounty.

The reward offer had been made by John Trott, the farmer who had plowed up the first underground orchid. Trott remained a keen naturalist throughout his life and had even named his farm Rhizanthella. Unfortunately he would die just a few months prior to the rediscovery of his "favorite species," and the reward would be honored by his widow.

On May 25, 1979, farmer John McGuiness of the Western Australian town of Munglinup, accidentally kicked over the dead root of a mallee eucalypt while clearing a section of natural bushland to use as a sheep pasture and found an orchid that had been playing hide-and-seek for two decades. After placing the carefully dug plant in a plastic cup, McGuiness drove the 350 miles to Perth and presented his flower to the botanical garden.

Alex George visited the site at Munglinup and helped to uncover an entire colony of *Rhizanthella gardneri*, a bonanza of thirteen orchids in flower. The population grows more than two hundred miles from the sites where other *Rhizanthella* plants have been discovered and collected.

Western Australian botanists are making the pilgrimage to the Munglinup site to clarify their knowledge of the orchid's life cycle. McGuiness has fenced off the site, and the botanists disturb the colony as little as possible. Five more localities were found over the next three years. A total of 150 flowering plants of this underground orchid have now been observed and carefully tagged, a time-consuming and uncomfortable process. Orchid hunters must crouch under thickets of broom honeymyrtle and daintily scrape away at the accumulated twigs and sand using gardener's forks with curved prongs similar to European truffle rakes.

Research on *Rhizanthella gardneri* is an example of what I like to call quantum-leap biology. A unique species is discovered, excites the interest of the scientific community, but basic knowledge of the organism soon languishes due to the paucity or low accessibility of specimens. The trickle of initial publications dries up quickly, and later papers consist of overgeneralizations and conjectures that add little new information to the particle of original evidence. Suddenly the species becomes available again, and professional investigations occur so rapidly and are so sophisticated that more is learned about the organism in two years than in the previous half century since its discovery. Quantum-leap biology is almost a way of life in Australia, considering the inverse proportion of the size of the human population versus the physical dimensions of the island continent. The community of professional biologists is just too small to cope with the full scope of natural diversity. While there are many clubs for competent and enthusiastic naturalists, amateurs usually lack the time and money to devote themselves exclusively to the rarer species. Underground orchids and marsupials, like the dibblers of chapter 6, comprise only a fraction of the Australian biota that have played peekaboo with generations of investigators.

A population of 150 sexually mature plants places *R. gardneri* within the endangered category, but since these individuals have been mapped and monitored, careful research can be nondestructive, yet illuminating. I have already mentioned advances and corrections in the underground orchid/fungus story. Observation of the Munglinup colony and plants at other sites have clarified some rather basic questions involving the presentation of the floral cup.

Rhizanthella really does flower underground, but rather close to the soil surface. As the creamy-white bracts pull away from the nest of flower buds they cover, they also push the thin, covering layer of dirt, fallen leaves, and grass straws upward. As the bracts age, they turn a deep mauve and continue to push up enough earth and detritus so that a "fairy mound" may be produced. In some cases the old bracts expand so far that the mound shifts and the flowers are exposed briefly to the open air through spaces between the partially uncovered bracts. Of course the small flowers always remain less than ¼ inch beneath the soil's surface even if their protective bracts are no longer covered with dirt and debris.

The underground orchid has two odors. If a portion of the rhizome is cut or bruised, the wound produces a strong stink reminiscent of formalin. Dr. George has suggested that this may be a protective device for such a soft, brittle plant. The little flowers also secrete a fragrance, but it is rather sweet and pleasant. Generally speaking, floral odors signal prospective pollinators that a food source is available. These flowers lack nectar glands, however, so some sort of pollination by deceit might be expected. Since it has been observed that the flowers of *Rhizanthella* do not self-pollinate with age, we must wonder what animal would pollinate the flower.

When I first read about the flowering habit of *Rhizanthella* and its colors and odor, I thought that beetles would be the most likely pollinators. Cream tinged with purple is the color of beetle-pollinated flowers, such as magnolias or the water lilies I discussed in

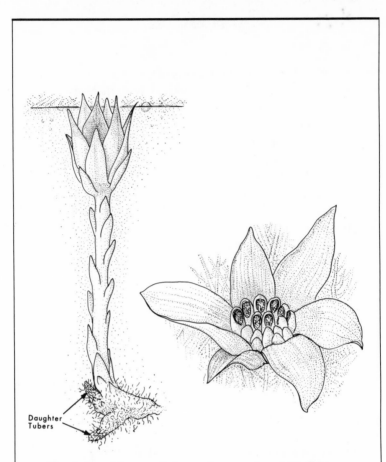

Daughter
Tubers

The hairy rhizome of the underground orchid (Rhizan-
thella gardneri) gives rise to one elaborate flowering stalk
and one or two daughter tubers when it is ready to repro-
duce. After the scaly stalk sets fruit and dies, the daughter
tubers continue to produce the next portions of rhizome. At
right, we see how the small, hooded flowers are nestled
within the cup of bracts as the flowering head begins to
expand just under the soil's crust.
Illustrations by J. Meyers.

chapter 9. Beetle flowers also smell sweet, as they mimic the odors of ripe, fermenting fruit, also appealing to insect scavengers. Some of the Australian leek orchids *(Prasophyllum)* are pollinated, in part, by beetles, so I suggested that *Rhizanthella* exhibited a more specialized form of presentation and would probably attract crawling insects like the rove beetles (Staphylinidae). I became guilty of the same overgeneralizing as other scientists who make hypotheses regarding the biology of species they've never seen. Both beetles and ants have been observed entering the cup of bracts, but they emerge without carrying wads of pollen (pollinia).

On June 23, 1980, John Cooke, a wildlife photographer, captured a tiny fly leaving the cup of a *Rhizanthella* orchid at the Munglinup site. The fly was carrying a wad of *Rhizanthella* pollinia on its back and was identified as a species of *Megaselia* in the family Phoridae. The phorid gnats have not been well studied, but their maggots grow in fungi or in rotting plant material that is infested with fungal threads. As adult, winged insects, some phorids are known to eat pollen as a protein source, and at least one Central American species is the most important pollinator of *Theobroma*, the bush whose seeds make chocolate.

The discovery of the *Megaselia* gnat may clarify a few mysteries about the sex life of *Rhizanthella*. This is a winter-flowering orchid, which means that it begins to bloom during the wet, cool weather at the same time as the helmet orchids *(Corybas)*, dead horse orchids *(Acianthus)*, and greenhoods *(Pterostylis)*. These three genera exploit the flies that are dependent on winter mushrooms. As *Rhizanthella* is so often associated with decaying branches and detritus, its flowering head emerges at the sites where phorid flies probably lay their eggs. Of course, we still cannot say whether the flowering head mimics a mushroom or if the sweet floral odor is offering the false promise of nectar and edible pollen. There have been some rumors that botanists have observed termites entering the capitulum and carrying off the pollinia! If this is true, and I'll have to see it to believe it, then the flowering head might be a compound trap mimicking fungal infestations.

The fertilized ovaries have their own mysteries to conceal. At one time scientists believed that the old capitulum broke through the soil surface, and ripened capsules released their seeds on the ground breezes. The ovaries of *Rhizanthella*, though, do not mature to form dry, self-opening capsules. They swell up just a little bit and become fleshy berries. The seeds, however, are smaller than poppy seeds and wear a hard, blackish shell, which may explain why it takes weeks for them to germinate. I have seen *Rhizanthella* seeds grow in flasks at the orchid lab at the National Botanic Garden in Canberra. *Rhizanthella* is now one of only four genera of orchids known to grow both fleshy fruits and crusty seeds. This elite club includes the reed orchids *(Apostasia)*, South American slipper *(Selenipedium)*, and the vanilla orchids *(Vanilla)*.

Seeds surrounded by juicy tissue suggests that *Rhizanthella* is dispersed by animals, which eat the berries, then defecate the seeds. Alex George suggests that *Rhizanthella* berries might be eaten by marsupials such as woylies *(Bettongia penicillata)*, rat-sized, hopping animals that also dig for underground mushrooms.

We can see that the mere rediscovery of *Rhizanthella* does not solve all the riddles of the survival of underground orchids. Hyacinth orchids and broom honeymyrtles are found throughout the southern half of Australia, so why are the ranges of the two underground species so restricted? Is there some growth requirement we do not understand yet or are these subterranean plants more common than we think? Is Rupp's *Cryptanthemis* gone forever or is this species waiting for its own rediscovery?

These orchids were discovered by pure chance. To find them by design would mean investigating huge tracts of land. It's likely that these plants are mere precursors to future discoveries. The earth conceals many more tantalizing stories of botany. Will there be enough time and qualified people to tell them?

Annotated Bibliography

Chapter 1

JANZEN, DANIEL H. *Ecology of Plants in the Tropics.* London: Edward Arnold, 1975. This paperback is extremely deceptive. It is brief and written in a style accessible to most bright high school students. Within less than seventy pages Janzen's outline presents one of the most informative primers on the interaction between tropical trees and their environments, emphasizing interdependency between plant and animal.

JANZEN, DANIEL, H. (editor). *Costa Rican Natural History.* Chicago: University of Chicago Press, 1983. Anyone expressing an interest in the New World tropics must read this book. Life zones within a tropical country are described in depth, and there is a most interesting treatment of the history of natural history in Costa Rica. The chief attraction of this book, though, is the meticulously prepared portraits of the life cycles of native species. This is one of the few treatments of a country's biota in which plants do not take a backseat to the animal life.

*MARSHALL, LARRY, G. "Land Mammals and the Great American Interchange." *American Scientist* 76(1988): 380–88. Central America was the landbridge by which the mammals of North America once colonized South America. This article provides a timetable of events regarding the arrival, mixing, and extinction of the largest creatures that once lived in the tropical lowland forests.

*WITTSBERGER, DENNIS; CURRENT, DEAN; and ARCHER, EDGAR. *Arboles del Parque Deininger* San Salvador, El Salvador: Ministerio de Educacion, Direccion de Publicaciones, 1982. This represents a triumph for Peace Corps volunteers assigned to work as foresters in one of the first national parks in El Salvador. They found 144 tree species native to the park. Each species receives its own illustration and description, including precious records of leaf, flower, and fruiting cycles as well as both local and commercial uses. Publication was delayed for years due to obvious problems, but we must applaud the courage of the few people who completed this project.

Chapter 2

GIBBS, MAY. *The Complete Adventures of Snugglepot and Cuddlepie.* Sydney, Aus.: Angus & Robertson Publishers, 1946. The book has never been out of print. Angus & Robertson now offers a paperback edition through its Bluegum paperback series. While the book is difficult to locate in the United States, botanical gardens and other institutions that are members of the Garden Society have started selling editions in their gift shops.

MULLINS, BARBARA. *Australian Banksias.* Sydney, Aus.: A. H. & W. Reed Pty Ltd., 1970. A very small but excellent introduction to the genus, which captures the range of growth forms, flowering and fruiting cobs, and the role the plants play in the lives of animals and humans.

PARKER, K. L. (collector). *Australian Legendary Tales.* Sydney, Aus.: The Bodley Head Ltd., 1978. Mrs. Parker collected aboriginal myths and published the first volumes in 1896. There are many more collections of aboriginal myths, but this is among the easiest to find. Aboriginals had completely different interpretations of their flora and fauna, with violent creation stories that Andrew Lang compared to the early Greek myths.

WICK, PETER A. *The Court of Flora (Les Fleurs Animees): The Engraved Illustrations of J. J. Grandville.* New York: George Braziller, Inc., 1981. Here are some of the elegant ancestresses of Little Ragged Blossom. Illustrating a series of stories that were published in 1847, Grandville took the fashions of his day and turned them into costumes for the spirits of flowers that left the celestial garden to live as women. Grandville takes basic floral anatomy and turns it into a series of witty visual puns.

Chapter 3

*CALDER, MALCOLM; and BERNHARDT, PETER (editors). *The Biology of Mistletoes.* Sydney, Aus.: Academic Press, 1983. Eighteen authors discuss their specialty. The chapters range from a treatment of mistletoes and continental drift to DNA variation. Australian species are emphasized, but there is plenty of information on the mistletoes of Europe, Africa, and the United States. The opening chapter, "Mistletoes in Focus: An Introduction," by Malcolm Calder (my former adviser and mentor), puts the plants in perspective for the novice loranthophile.

KUIJT, JOB. *The Biology of Parasitic Flowering Plants.* Berkeley: University of California Press, 1969. Kuijt discusses all the plant families that make flowers and haustoria. Mistletoes share the book with the boomrapes, dodders, rafflesias (queen of the parasites), beech drops, and many other vampires. Profusely illustrated with careful line drawings, this book affords an exciting introduction to the evolution and life cycles of parasites from American deserts to Malaysian rain forests.

Chapter 4

GREAT PLAINS FLORA ASSOCIATION. *Flora of the Great Plains.* Lawrence, Kans.: University Press of Kansas, 1986. We finally have a definitive account of every plant species currently known to inhabit the American prairies. The descriptions take up almost fourteen hundred pages.

KNOX, R. B. *Pollen and Allergy. Studies in Biology,* no. 107. London: Edward Arnold Publishers Limited, 1979. The biochemistry and aerodynamics of hay fever are presented in such a clear manner that it will make your own discomfort more comprehensible, if not bearable.

REICHMAN, O. J. *Konza Prairie: A Tallgrass Natural History.* Lawrence, Kans.: University Press of Kansas, 1987. Reichman examines the development of the prairie over time and the seasonal interactions between plants and animals. There is special emphasis on the different ecosystems that make up one whole prairie.

Chapter 5

*ESAU, KATHERINE. *Anatomy of Seed Plants,* 2nd ed. New York: John Wiley and Sons, 1977. Once upon a time, not too long ago, courses on plant anatomy and morphology were offered by most biology departments at American universities. This was one of the most important texts for such a course. Esau discusses the development of flower organs in the bud to help understand the variation in floral symmetry.

*FAEGRI, KNUT; and LEENDERT, VAN DER PIJL. *The Principles of Pollination Ecology,* 3rd ed. New York: Pergamon Press, 1977. This classic remains accessible, since it has been revised as needed and remains easy to read.

MEEUSE, BASTIAAN; and MORRIS, SEAN. *The Sex Life of Flowers.* New York: Facts-on-File Publications, 1984. The prose has been criticized for being too "geewhiz," and there may be some justification for this. However, there is no other book with finer photography depicting the range of floral forms and the animals associated with particular forms.

Chapter 6

*HOPPER, STEPHEN D. "Bird and Mammal Pollen Vectors in Banskia Communities of Cheyne Beach, Western Australia." *Australian Journal of Botany* 28(1980): 61–75. Steve's compared honeyeaters with honey possums. The latter does not appear to be as efficient a pollinator.

SUSSMAN, R. W.; and RAVEN, PETER H. "Pollination by Lemurs and Marsupials: An Archaic Co-evolutionary System." *Science* 200(1978): 731–34. Plate tectonics, paleontology, and modern observations are combined in this short review to develop a new theory. To make sense of the diversity of life, a modern evolutionary biologist must become familiar with many fields of study.

TERBORGH, JOHN; and STERN, MARGARET. "The Surreptitious Life of the Saddle-backed Tamarin." *American Scientist* 75(1987): 260–69. The life of small primates involves close family units and large territories. The authors give a good account of environmental pressures and food restrictions.

*TURNER, VIVIENNE. "Marsupials as Pollinators in Australia." pp. 55–66. In *Pollination and Evolution,* edited by J. A. Armstrong, J. M. Powell, and A. J. Richards. Sydney, Aus.: Royal Botanic Gardens, 1982. This article is derived from a seminar Vivienne presented before the International Congress in 1981. The audience was so unfamiliar with the native marsupials at the time that they began to giggle at Viv's slides. This article does provide a fine assessment of pollen- and nectar-feeding habits and introduces Viv's interpretation of the plant life's long-term response to possums and parrots.

Chapter 7

HAREUVENI, NOGAH. *Nature in Our Biblical Heritage.* Translated by Helen Frenkley. Kiryat Ono, Israel: Neot Kedumim Ltd., 1980. Things are not always as they seem. Familiar customs and religious objects often draw their inspiration from plants native to the Middle East.

HAWKESWOOD, TREVOR. *Beetles of Australia.* Sydney, Aus.: Angus & Robertson Publishers, 1987. There do not appear to be any semitechnical books in English on the lives of beetles in the Mediterranean regions. However, Hawkeswood's introduction provides both an excellent text and a collection of photographs of flower-visiting beetles in *Coleoptera*'s major families, from the Old World and Australia in particular.

ZOHARY, MICHAEL. *Plants of the Bible.* Cambridge, Eng.: Cambridge University Press, 1982. There is a long tradition of Bible botany, and these are among the most recent entries directed toward the interests of gardeners and naturalists. Zohary provides sketches on ecology and life history in addition to the usual translations of pertinent passages.

Chapter 8

*RAVEN, PETER H.; and AXELROD, DANIEL I. "Angiosperm Biogeography and Past Continental Movements." *Annals of the Missouri Botanical Garden* 61(1981): 539–673. The birth and spread of flowering plants is interpreted on the bases of the fossil record and plate tectonics. The authors show how a number of plant relicts have come to live in isolated "retirement villas," especially around the tropical belt.

*THIEN, LEONARD B.; et al. "The Pollination of *Zygogynum* (Winteraceae) by a Moth, *Sabatinca* (Micropterygidae): An Ancient Association." *Science* 227(1985): 540–43. Committee papers may upset purists, but there is some merit to letting specialists concentrate on different aspects of the same problem. Gordon McPherson located and identified the trees in New Caledonia. George Gibbs identified our moths, while Leonard Thien completed all the fieldwork. I examined the pollen and showed how it became attached to the moths, while the remaining authors analyzed the fragrances.

*WALKER, J. W.; BRENNER, G. J.; and WALKER, A. G. "Winteraceous Pollen in the Lower Cretaceous of Israel: Early Evidence of a Magnolialean Angiosperm Family." *Science* 220(1983): 127. The structure of the wall of the extinct pollen cluster is described and then compared to pollen walls of extant Winteraceae.

Chapter 9

COATS, PETER. *Flowers in History.* New York: Viking Press, 1970. A chapter on the cultivated water lilies describes the history of *Victoria* in England, offering lots of memorabilia concerning the first blooming of the giant plant.

*PRANCE, GHILLEAN T.; and ARIAS, JORGE R. "A Study of the Floral Biology of *Victoria amazonica* (Poepp.) Sowerby (Nymphaeaceae)." *Acta Amazonica* 5(1975): 109–139. It is most unlikely that this paper will be surpassed in the near future, for it is one of the best accounts of the reproductive cycle of any tropical plant known.

Chapter 10

BAIRD, VIOLA BRAINERD. *Wild Violets of North America.* Berkeley: University of California Press, 1942. There have been more recent revisions of the genus *Viola* in the United States, but this one was written by a second-generation violaphile, who inherited her father's eye for detail. The illustrations are both meticulous and charming.

*BEATTIE, A. J. "Plant-Animal Interactions Affecting Gene Flow in Viola." In *The Pollination of Flowers by Insects,* edited by A. J. Richards. Linnean Society Symposium Series no. 6, pp. 151–64. London: Academic Press, 1978. This chapter sums up Andrew's work on reproduction in the violets of Britain and the

United States, describing his fieldwork from the late sixties through the early seventies.

*BEATTIE, A. J. "Floral Evolution in Viola." *Annals of the Missouri Botanical Garden* 61(1974): 780–93. Why do some violets wear nectar guides on each petal and others have the guides confined discretely to the spur petal? This paper offers important thoughts on the evolutionary pathways that have painted and sculpted flowers on different continents.

COON, NELSON; and GIFFEN, GEORGIANNE. *The Complete Book of Violets.* South Brunswick, N.J.: A. S. Barnes and Co., Inc.; 1977. Written by a man who was probably the last individual to grow violets commercially on the east coast of the United States. Despite the title, the book contains little biology but brims with violet poetry, recipes, history, and folklore. Beware, though, of Coon's well-entrenched belief in his favorite plants' abilities to cure serious illness.

Chapter 11

*BANKS, JO ANN. "The Reproductive Biology of *Erythronium propullans* Gray and Sympatric Populations of *E. albidum Nutt.* (Liliaceae)." *Bulletin of the Torrey Botanical Club* 107(1980): 181–88. An exceptional contribution to trout lily biology that compares a species that has "given up" sex to a common, "sexually active" population. This paper demonstrates how scientists can perform simple experiments in the field to gather a lot of information.

*BERNHARDT, PETER. "The Pollination Ecology of a Population of *Erythronium americanum Ker.* (Liliaceae)." *Rhodora* 79(1977):

278–82. An inconsequential work, if I do say so myself, but it was only the second paper I published. Its only merit is that it seems to have stimulated greater interest in the reproductive biology of *Erythronium.*

*KAWANO, SHOICHI; HIRATSUKA, AKIRA; and KAZUHIKO, HAYASHI. "Life History and Characteristics and Survivorship of *Erythronium japonicum.* "*Oikos* 38(1982): 129–49. A cradle-to-grave assessment of the life of trout lily populations in Japan. This holistic treatment of a species has become a sophisticated tool of population biologists. You may find it interesting to see how a wildflower can be treated as a model system.

RICKETT, H. W. *Wildflowers of the United States.* New York: McGraw-Hill Books Co., 1965. Offers a look at the distribution of trout lilies and their relatives across this country. I still consult the volumes when I am about to take a trip to an unfamiliar part of America.

*STEBBINS, G. L. *Flowering Plants: Evolution Above the Species Level.* London: Edward Arnold Publishers, 1974. Ledyard Stebbins remains one of our most provocative authorities on evolutionary trends in the seed plants. He comments on droppers and explores survival mechanisms in lilylike plants.

Chapter 12

BENZING, DAVID H. *The Biology of the Bromeliads.* Eureka, Calif.: Mad River Press Inc., 1980. Although this book deals with all aspects of the life cycles of bromeliads and man's influence on the Bromeliaceae, Benzing offers a synopsis of most of his earlier research. He is most adept at describing the physiology of water transport in different growth forms and how microscopic structures accommodate environmental changes.

*PITTENDRIGH, C. S. "The Bromeliad-Anopheles-Malaria Complex in Trinidad. I—The Bromeliad Flora." *Evolution* 2(1948): 58–89. A classic paper best appreciated as a scientific detective story. Later research has shown many more levels of dependency between tropical animals and tank bromeliads.

STEELE, ARTHUR R. *Flowers for the King.* Durham, N.C.: Duke University, 1969. The reader is offered a glimpse of the new science of botany during the rise and fall of the Spanish empire. Bromeliads and many other plants of the New World tropics were scrutinized and exploited during the age of the conquistadors.

Chapter 13

*BURNS-BALOGH, PAMELA; and HESSE, MICHAEL. "Pollen Morphology of the Cypripedioid Orchids." *Plant Systematics and Evolution* 158(1988): 165–82. *Cypripedium* and its allies are examined from quite a different angle, taking the reader away from pretty flowers and permitting an in-depth exploration of the sponge-like wall of the pollen grain. Studies like this one help scientists compare *Cypripedium* to more advanced orchids and to nonorchidlike plants.

CORRELL, D. S. *Native Orchids of North America North of Mexico.* Waltham, Mass.: Chronica Botanica Company, 1950. This book remains a classic on the natural history and taxonomy of the orchids of the United States and Canada. It has quite a lot of information on *Cypripedium* species which is unavailable in more recent books.

LUER, CARLYLE A. *The Native Orchids of the United States and Canada Excluding Florida.* New York: The New York Botanical Garden; Ipswich, Eng.: W. S. Cowell Ltd., 1975. Widely regarded as the modern, definitive study of North American orchids, this volume is especially useful for its careful and expansive study of the *Cypripedium* species. No one who claims any sort of affection for slipper orchids can fail to be impressed by the color photography, which will actually help acquaint the reader with the growth habits and natural variation of his favorite species.

RANDOLPH, VANCE. *Ozark Magic and Folklore.* New York: Dover Publications, 1947. This paperback edition, first published in 1964, is an unabridged republication of Randolph's original work, which was published by Columbia University Press. I found this an invaluable introduction to plant folk medicines and superstitions. Despite the blandishments of the holistic-medicine proponents, Randolph shows just how uncomfortable and nasty these treatments can be.

Chapter 14

*CATLING, PAUL H.; BROWNELL, V. R.; and LEFKOVITCH, L. P. "Epiphytic Orchids in a Belizean Grapefruit Orchard: Distribution, Colonization, and Association." *Lindleyana* 1(1986): 194–202. An exciting and modern contribution to the study of orchid dispersal and survival. This is a careful assessment of what happens when man abandons an agricultural site in the tropics and how the epiphytic orchids move in. There is a fascinating correlation between citrus-branch dimensions and where different orchid species will survive over time.

DRESSLER, ROBERT L. *The Orchids: Natural History and Classification.* Cambridge, Mass.: Harvard University Press, 1981. This may be the best book on orchid biology available in English. Dressler writes with the competence of a field naturalist with almost thirty years of experience in Central and South America. He is especially lucid on matters of epiphyte ecology and survival. The illustrations and photographs are numerous and clarify all major concepts. Microscopic seeds are well depicted.

HAMER, FRITZ. *Las Orquídeas de El Salvador.* Volumes 1 and 2. San Salvador, El Salvador: Ministerio de Educación, 1974. Don't let the Spanish title scare you away. The text is written in Spanish, English, and German. Hamer discusses the distribution of the species in relation to climate and altitude. Each species is illustrated, and many receive their own color photograph.

LUER, CARLYLE A. *The Native Orchids of Florida.* New York Botanical Garden. Ipswich, Eng.: W. S. Cowell Ltd., 1972. Several urban orchids, discussed in this chapter, have wide distributions— Florida marks their most northern range. Luer's fine descriptions and color photography encourage orchid watching without taking a trip to a Latin American city.

Chapter 15

BOYLE, F. *About Orchids—A Chat.* London: Chapman and Hall, 1893. Boyle was one of the great publicists for tropical-plant businesses in his day. He was fond of "purple prose," describing the orchid hunter in his jungle element. It's not easy to locate this book, which is too bad, since Boyle offers a glimpse through a rose-colored window on the past, whether attending an orchid auction or recording the gossip and the folklore of a hobby.

REINIKKA, MERLE. *A History of the Orchid.* Coral Gables, Fla.: University of Miami Press, 1972. Orchid history is presented as a series of overlapping biographies. Both knowledge and profit have pursued the orchid muse over several centuries.

TYLER-WHITTLE, M. S. *The Plant Hunters.* Philadelphia: Chilton Book Co., 1970. Tyler reflects on how expeditions for plants and the personalities involved in such major projects have influenced both horticultural trends and the young science of botany. The phenomenon of orchid fever is much more easily understood when Tyler places it in its historical context as just one link in a chain of fads revolving around the acquisition of tropical curiosities.

Chapter 16

HILLERMAN, FRED E.; and HOLST, ARTHUR W. *An Introduction to the Cultivated Angraecoid Orchids of Madagascar.* Portland, Oreg.: Timber Press, 1986. This is the only modern English treatment of the most popular comet orchids and their relatives.

*NILSSON, L. ANDERS. "The Evolution of Flowers with Deep Corolla Tubes." *Nature* 334(1988): 147–49. The "direction" of spur evolution is investigated, giving live sphinx moths access to flowers with spurs shortened artificially.

*NILSSON, L. ANDERS; et al. "Angraecoid Orchids and Hawkmoths of Central Madagascar: Specialized Pollination Systems and Generalist Foragers." *Biotropica* 19(1987): 310–18. There is strong support here for the sort of scientific research that does not require a lot of ponderous and expensive equipment. Two weeks spent in a rather decimated forest have changed concepts regarding the pollination of tropical orchids.

Chapter 17

*LUER, C. *Thesaurus Dracularum: A Monograph of the Genus Dracula.* St. Louis: Missouri Botanical Garden, 1988. When the taxonomy of a genus of plants is established or revised, it is treated to a full, scientific monograph. Luer employs a fine artist and careful descriptions to produce a lavish but accurate spook show. *Dracula* species are believed to be pollinated by fungus gnats.

WELLS, H. G. *The Stolen Bacillus.* London: Methuen, 1895. This is the anthology that first housed Wells's orchid story. It's always worthwhile reading his collections as Wells offers a running critique on the pursuits of naturalists during the late Victorian and Edwardian periods. Bird stuffers, butterfly hunters, and the collectors of "natural curiosities" are also parodied. Plants and plantlike organisms play pivotal roles in much later stories, such as "The Apple," "The Purple Pileus," and "A Slip Under the Microscope."

Chapter 18

*DIXON, K. W.; and PATE, J. S. "Biology and Distributional Status of *Rhizanthella gardneri Rogers:* The Western Australian Underground Orchid." Kings Park Research Notes, no. 9(1984). How do you locate and sample a rare species without making it even rarer? Most of this report is a rather dull account of the vegetation associated with the underground orchid, but the authors include tidbits on history and environmental pressures not available in earlier papers.

GEORGE, ALEXANDER S. *"Rhizanthella gardneri R.S. Rogers*—The Underground Orchid of Western Australia." *American Orchid Society Bulletin* 49(1980): 631–46. A successful attempt to marry the earliest papers with the sudden rush of new information.

*WARCUP, J. H. "Mycorrhizas and Growth of *Rhizanthella gardneri.*" *New Phytologist* 99(1985): 273–303. No serious student of orchid conservation can afford to miss Warcup's ongoing research. This study, in particular, shows the problems inherent in isolating orchid fungi and testing them on the next generation of seeds.

*Denotes scientific literature that is probably unavailable in most public libraries. Readers are encouraged to look in the libraries of botanical gardens and arboreta, natural history museums, and any university or college with its own department of biology or botany.

Index